BRAIDING FINE LEATHER

BRAIDING FINE LEATHER

Techniques of the Australian Whipmakers

David W. Morgan

Cornell Maritime Press
A Division of Schiffer Publishing, Ltd.
Atglen, Pennsylvania

Published by Schiffer Publishing Ltd.

Braiding Fine Leather: Techniques of the Australian Whipmakers was originally publised by Cornell Maritime Press, Inc. in 2002

4880 Lower Valley Road
Atglen, PA 19310
Phone: (610) 593-1777; Fax: (610) 593-2002
E-mail: Info@schifferbooks.com

For the largest selection of fine reference books on this and related subjects, please visit our web site at **www.schifferbooks.com**
We are always looking for people to write books on new and related subjects. If you have an idea for a book please contact us at the above address.

This book may be purchased from the publisher.
Include $5.00 for shipping.
Please try your bookstore first.
You may write for a free catalog.

In Europe, Schiffer books are distributed by
Bushwood Books
6 Marksbury Ave.
Kew Gardens
Surrey TW9 4JF England
Phone: 44 (0) 20 8392 8585; Fax: 44 (0) 20 8392 9876
E-mail: info@bushwoodbooks.co.uk
Website: www.bushwoodbooks.co.uk

Library of Congress Cataloging-in-Publication Data

Morgan, David W., 1925-
Braiding fine leather : techniques of the Australian whipmakers / David W. Morgan.
p. cm.
Includes bibliographical references.
ISBN 978-0-87033-544-0
1. Leatherwork. 2. Braid. I. Title.
TT290 .M57 2002
685—dc21

2002012078

Printed in China
First edition; second printing 2005, fifth printing 2010

This book is dedicated to those generations of Victorian craftsmen who went before us and left an inspiring legacy of high standards and a spirit of innovation.

George Vogt, Sydney, Australia, 1968

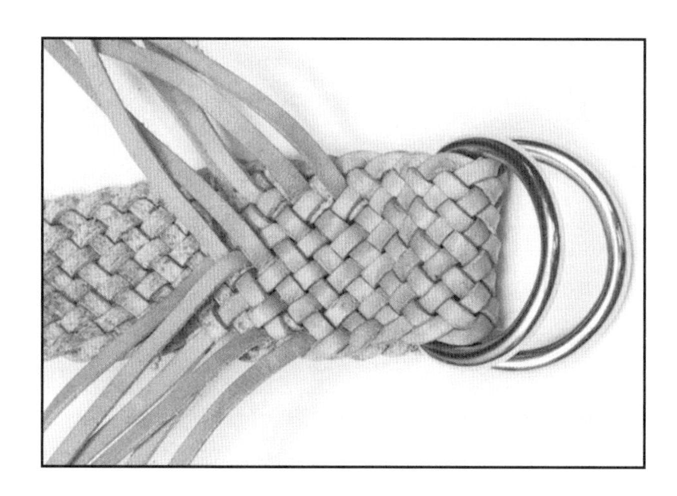

Table of Contents

Preface

Leather braiding is a richly satisfying craft. The end products may range from strictly utilitarian items such as cow halters to elegant items of apparel such as hatbands. Accompanying this variety of applications is a range of difficulty in making the items. For many utilitarian items, the cutting, braiding, and finishing processes are not difficult, and these are readily handled by the novice. As a braider gains skill and confidence, the choices in design, leather, cutting, braiding, and finishing reflect the wider scope available to someone with suitable experience. The highest levels of craftsmanship, such as those found in the best quality whips and horse tack, are restricted to those few professional braiders who can combine the disciplined experience of a working craftsman with artistic skill and the drive of a perfectionist. However, the amateur braider can produce good work and gain full satisfaction at every level of competence.

This book is designed to help the beginning leather braider acquire the needed basic skills in a straightforward manner, so that simple braiding projects can be done well at an early stage. With a little practice along the lines suggested and with close attention to detail, the beginner will soon be able to produce attractive and enduring items from either precut lace or from a skin or side of leather.

Use of rawhide is not covered in this book. However, the techniques used with rawhide are much the same as those used with leather, if allowance is made for the different properties of the material. The rawhide braider will find much pertinent information in the descriptions given in this book.

The techniques presented are largely drawn from Australian practice. Leather braiding was highly developed in Australia in whipmaking shops. The craft was carried to Australia by thongmakers trained in England, who were familiar with the thongs used on the finely braided carriage whips. Kangaroo leather, one of the finest leathers available for braiding, provided the material for high-quality work. A large and discriminating market supported many shops, large and small. Competitive conditions provided incentive for the refinement of techniques. New techniques, particularly those developed in the large shops, were quickly recognized and adopted throughout the trade. While whipmaking is no longer a major trade in Australia, it is still carried on, and the techniques developed in more harshly competitive times are still used or remembered. These techniques provided the main basis for this book.

The general discussion in the first part of the book gives an overall view of braiding and some explanation of the characteristics of braided work. This section is designed not only for the braider but also for the nonbraider who may use braided work, or for the leatherworker who may wish to extend his or her range of familiarity beyond work in solid hide. The basic techniques used in braiding are presented as a group, prior to the introduction of instructions for specific projects. It is recommended that the beginner experiment with various basic techniques, such as varying the width of strands, paring, and so on. While at first glance it may appear somewhat dull to make up a series of four-strand braid samples, much can be learned about quality work by such experimentation, and the more serious braider will find this preliminary work invaluable. Experimental pieces, with notes on the materials used, can also be helpful later in establishing the widths and lengths of lace to use in other projects. The projects in the second part of the book show the application of the basic techniques from the first part, and they also present selected secondary techniques, largely those used in the starting or finishing of pieces. The items produced are typical braided goods, and they cover a reasonably broad although not comprehensive range. Knots necessary to complete the projects are presented. Those wishing to explore knots further will find many sources in the extensive literature on the subject.

This book will provide a sufficient basis so that the braider will be able to handle most straightforward braided work—including the usual cowboy work of the American West—confidently and competently, whether reading directions from books or following samples of work made by others. Instructions for whips and fancy braiding have not been included, although the more adventurous braider will have little difficulty progressing to the fancy work on his own.

For further projects and a good coverage of American work and fancy knots, Bruce Grant's *Encyclopedia of Rawhide and Leather Braiding* is suggested; the book also contains an extensive bibliography. Those who wish to pursue tapered work and/or whipmaking will find some help in my earlier book, *Whips and Whipmaking,* and in *How to Make Whips* by Ron Edwards. Further references and sources for materials and tools are shown following the text.

The first draft of this book was largely written soon after the publication of *Whips and Whipmaking,* and I should like to reiterate my acknowledgment in that book to those who were so generous in their help. Through practical experience in the intervening years, I have gained a greater understanding of—and a deeper appreciation for—the skills developed by Australian, and before them, English, thongmakers. The experience gained by teaching others to braid has proven invaluable in clarifying the details involved in braiding well, and I should like to acknowledge the indirect contribution of all those whom I have taught.

I should like to thank my wife Dorothy, who made this book possible, initially by introducing me to Australia and later by helping me with the first draft. I also wish to acknowledge specific help from three capable braiders. My daughter Meredith encouraged me to complete this book. Without that encouragement the first draft would lie untouched. Her assistance in organizing the text has done much to bring the project to completion. My daughter Barbara assisted with the coordination of illustrations and text to ensure that they convey the information and procedures as intended. Meagan Thomas critically reviewed all projects and assisted with the photography.

BRAIDING FINE LEATHER

CHAPTER ONE

Background

TECHNICAL ASPECTS OF BRAIDING

Leather is one of the oldest materials used by man. From earliest times to the present day, leather's strength and suppleness, coupled with ready availability, have provided the basis for its selection over other materials for demanding applications. One use of leather has been for straps or ropes, which take advantage of the material's strength as well as the ease with which it can be cut into strips. A closely related use is braided leather—instead of the leather being cut into a single wide strap, it is cut into several narrow ones, which are then braided together to form one wide (or round) strap.

A typical round braid is a rope made from four or more strands, half the strands spiraling to the left and half to the right, each strand weaving under and over the strands in the opposite spiral in a regular pattern (fig. 1-1). There may or may not be a central core inside the braided strands. Such a braided rope has several advantages over a single strand strap of equivalent cross section. An understanding of these advantages is essential to any understanding of the development, design, or use of braided leatherwork.

Strength

The properties of leather are not uniform throughout a hide. Depending on the amount of movement associated with a particular portion of the animal, or the degree of protection a section may have, different parts of

Fig. 1-1. Eight-strand braid over core with two contrasting strands

the hide have different characteristics of strength and stretch. In addition, imperfections may be found, such as tick marks or flensing cuts—accidental cuts made when removing the skin from the animal. A single strand used as a strap is only as strong as its weakest section, but in a braided rope each strand supports the others; if a strand has an imperfection and breaks or stretches, the other strands will take up the load. A rope braided from four strands of leather with cuts in one or more strands could still have 75 percent of the strength of a rope braided of strands with no cuts, as long as no cuts overlap. Thus, braiding averages the strength of the leather and reduces the effect of flaws in the individual strands.

Flexure

The flexural stress on a braided leather rope is less than the stress on a single strand of leather of the same cross section. When a rope bends, the outer side of the bend is stretched, and the inner side is compressed. If the rope is made of a single strand, the stretch and compression must be taken up in full by the outer part of the strand. In a braided thong, the situation is quite different. Here, the individual strands follow a spiral path around the thong, and much of the stretch of the outer part of the thong (on the outside of a bend) is taken by separation of the strands. The individual strands are stressed during bending, but the maximum stress is far less than the maximum stress found in an equivalent single-strand thong. A Slinky, the coiled-wire toy, demonstrates this aspect of bending dramatically (fig. 1-2). The wire itself is stiff, and much more force is required to bend it than is required to bend the Slinky. In bend-

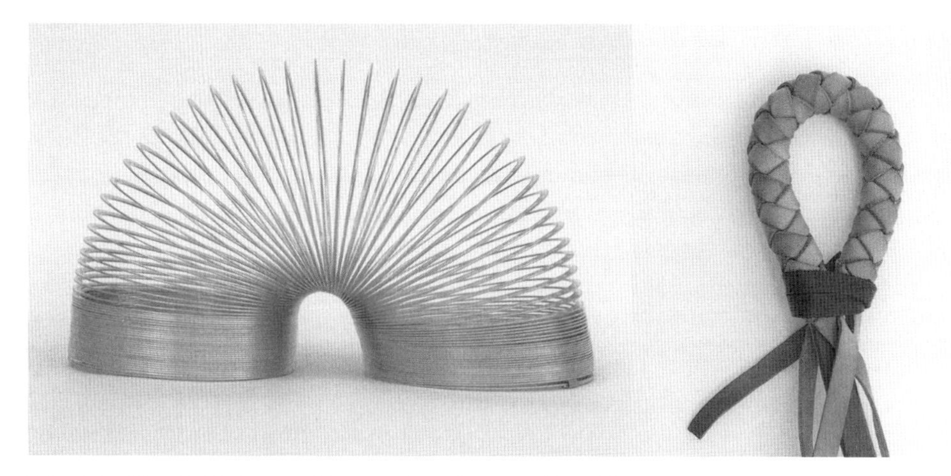

Fig. 1-2. Bent Slinky and four-strand round braid

ing the Slinky, the "stretch" on the outside of the bend is readily provided through separation of the individual coils; there is little stress on the wire making up the coils.

Size

A single strand of leather is limited in thickness to the thickness of the original leather. By braiding, straps or ropes can be made of much greater thickness or diameter than the thickness of the leather used (fig. 1-3).

Fig. 1-3. Four-strand round braid

Joins and Attachments

Because individual strands of leather may be spliced into a braid with no loss in strength, sections of braided work may be joined by back braiding, or interweaving the end strands of one braid into the body of its own braid or that of another braid. The braid can also be attached to hardware such as rings with no loss of strength (fig. 1-4).

Besides these four structural advantages over ordinary strap work, braiding provides different choices for both functional and decorative design considerations. From the standpoint of functional design, braided work, through the use of smooth spliced joins and variations in the width of strands, can tailor different parts in size and strength to meet specific needs. Braiding offers a variety of options for decorative design. Texture is controlled through the size, shape, paring (trimming corners of the strands), and number of strands. Coloring of the overall piece or individual strands can create products that range from the subdued to the flamboyant. Finally, covering knots and fringes can be used to transform a simple design into an ornate piece.

Good decorative design in braided work must start with good functional design. Much of the best decorative work comes from the refinement of essentially functional aspects. In California-style reins, for example, the knots positioned near the bit are both functional and deco-

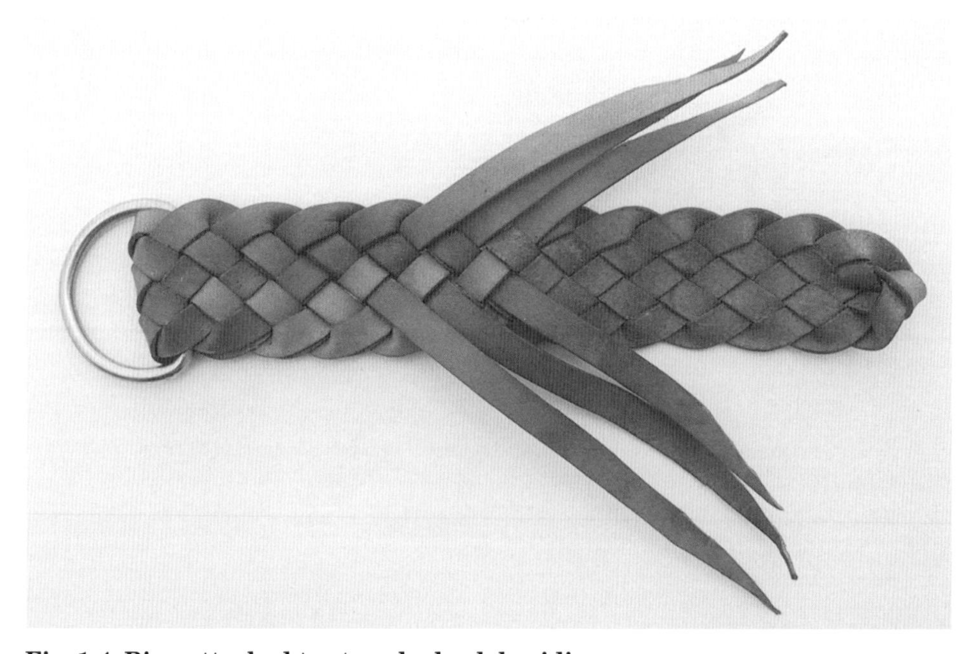

Fig. 1-4. Ring attached to strap by back braiding

rative. Other means could be used to develop the desired weight and flexure of the reins, such as chains or metal weights, but none serve better or are more attractive than well-made and well-designed knots.

HISTORICAL ASPECTS

Braided leatherwork is associated principally with the raising of livestock. Braiding developed as a craft to supply horse gear, ropes, and whips, particularly in areas where cattle are managed from horseback. The craft was initially practiced on an amateur or part-time basis. Seasonal work with cattle allowed time in the off-season for braiding. A few professional braiders supplied the needs of the cattlemen and horsemen, supplementing but not supplanting the occasional or seasonal workers.

The greatest development of braided leather or rawhide occurred where a sufficiently large and discriminating market was available to encourage professional braiders, and where economic or social conditions were such that men of considerable talent could be kept busy working in the trade. Those conditions prevailed in England in Victorian times and up to the First World War. The amateur carriage-driving enthusiasts from among the very rich took great pride in their equipment and were not only financially able but also sufficiently knowledgeable to demand the very best work. The professional whipmakers, working under hard but stable and productive economic conditions, produced the finest carriage whips ever seen.

Whipmaking in England evolved into two separate trades. Whipmakers specialized in making stocks or handles, buggy whips (known as straight-out whips), and riding crops. The flexible leather thongs to be attached to the stocks or handles were made by the second group of tradesmen, the thongmakers. The thongs were mounted on the stocks by the whipmakers. Whipmakers usually worked for larger shops catering to the retail trade. Thongmakers usually worked independently or in small groups, selling to the whipmaking establishments. Both trades worked with an apprenticeship system, ensuring good training and high standards, and for both the highest quality work was found in the carriage whip. The thongmakers also made the heavier thongs for cattle whips and foxhunting, although these were never refined to the same degree as the carriage thongs (fig. 1-5).

Immigrant English thongmakers formed the basis of the Australian whipmaking industry. The Australian whipmakers found their main trade in supplying whips for handling range cattle in the expanding beef

Fig. 1-5. English stock whip

industry, since carriage whips were imported from England. Working in kangaroo leather, they soon developed the Australian stock whip as a refinement of the English cattle and hunting whip. While the affluent market was small, the overall market was discriminating. Under the hard conditions prevailing from the late nineteenth century to the 1980s, Australian whipmakers produced stock whips of unsurpassed quality and beauty. Few fully trained whipmakers are still working in Australia, but numerous part-time and amateur whipmakers continue to produce stock whips of good quality.

Argentina still maintains the conditions of market and supply conducive to the development of braided horse gear. The braiders work largely in rawhide, often using foal skin. Their main production is in horse gear, reatas, and quirts, although a wide variety of whips, boleadoras, and fancy pieces such as tallies for counting (fig. 1-6) are made. The wealthy landowners provide a stable and discriminating market for the finest work, and the professional craftsmen work either under a paternalistic system on the large properties or under strongly competitive conditions in the cities.

Braiding has never become a strong commercial craft in the United States, despite a large market. Most of the commercial work sold in the United States has always been done in Mexico, where the braiders are in general too remote from their ultimate market to develop quality beyond minimal trade requirements. The domestic whip industry was mechanized when the lands in the West were being opened up to ranching and farming, and there was a large demand for whips. The wagon whips and shot-loaded team whips usually had sewn covers over the body of the whip, and only elementary skills were required to braid the

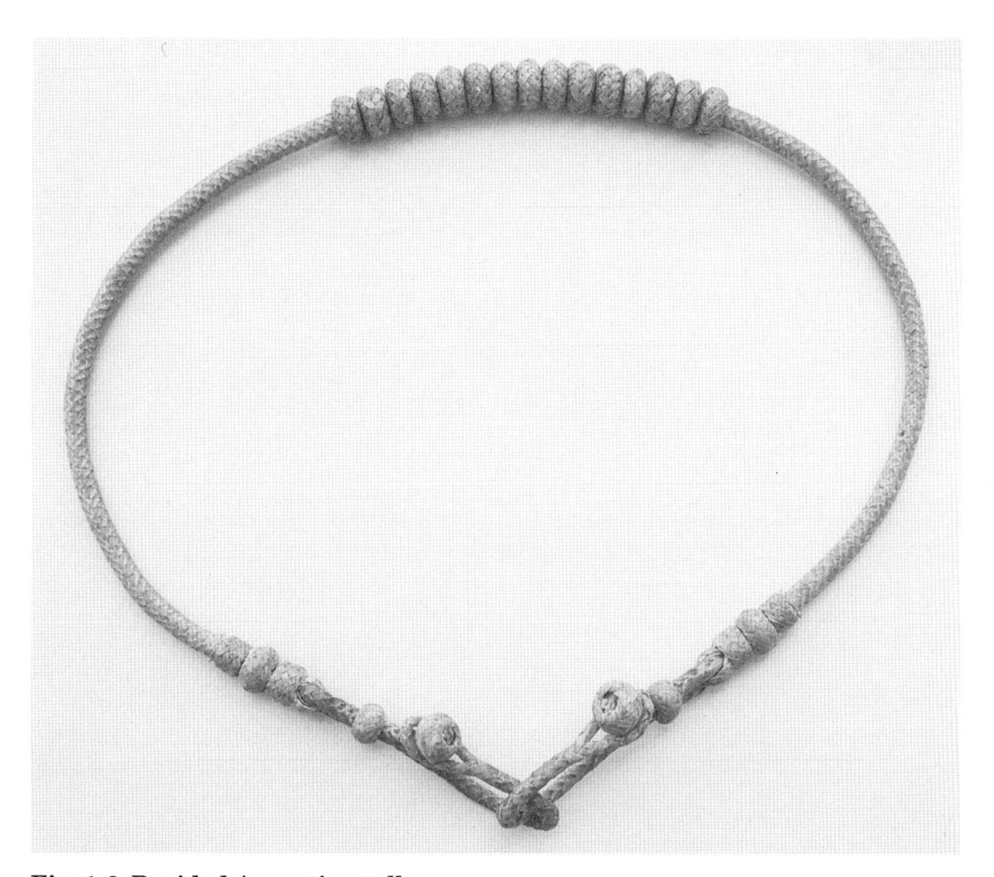

Fig. 1-6. Braided Argentine tally

point. The heavy bullwhips—which used the plentiful edges taken off belting leather to make the cores of the whips—were largely braided on machines. They were good whips, consistent in quality and inexpensive, but they never approached the quality of the Australian work. The best American braiding work has traditionally come from a few relatively isolated professional or amateur braiders, with the highest development possibly in California-style horse gear, primarily in rawhide but also in leather (fig. 1-7). Work of outstanding quality continues to be

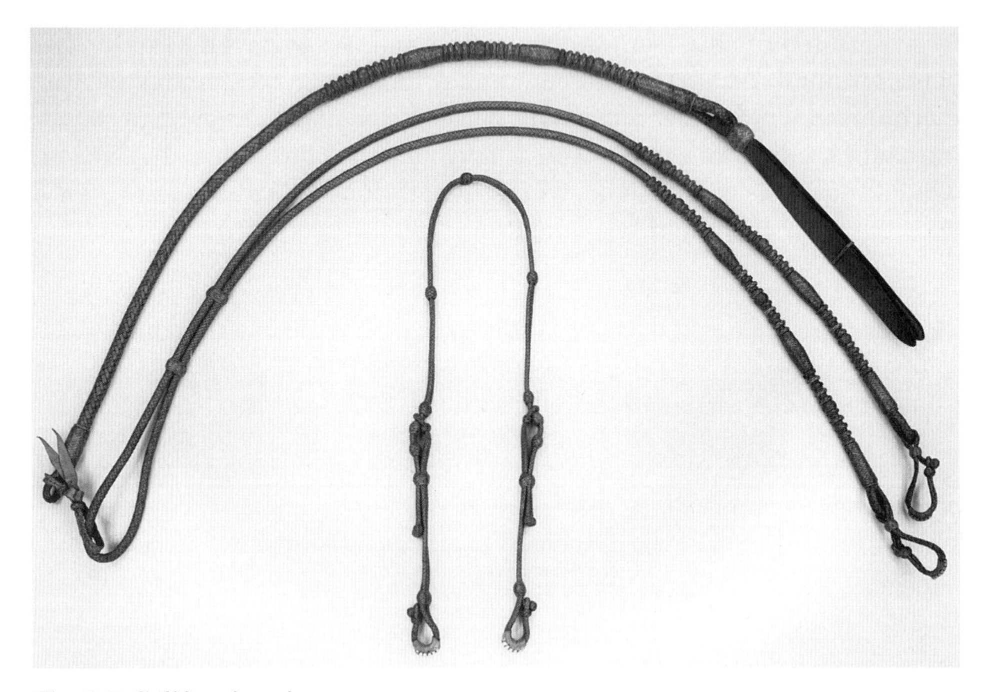

Fig. 1-7. California reins

made in small quantities and is both recognized and appreciated by horsemen throughout the United States.

CHAPTER TWO

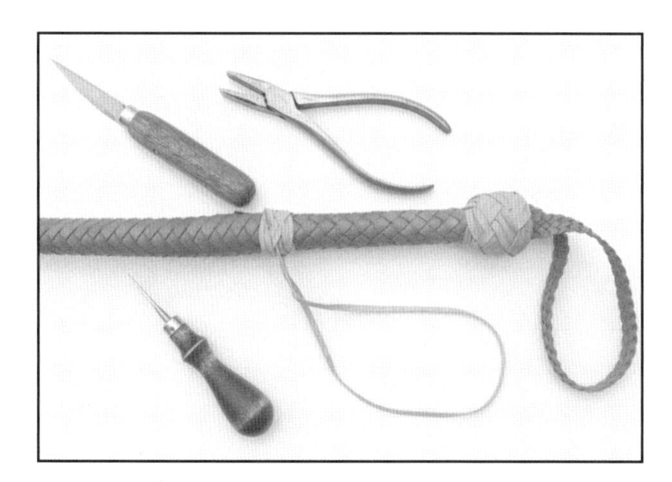

Tools

The tools required for braiding are simple and few, and little space is required for working. The following are the essential tools:

KNIFE

A good sharp knife is basic to any work in leather. A square-point leather knife or a pointed knife with a straight cutting edge may be used. Alternatively, a high-speed steel hacksaw blade may be ground to shape, with the back of the blade sharpened and the teeth removed for comfort in holding. On a narrow blade, the lower part may be wrapped with tape to form a handle. Knives with replaceable snap-off blades are satisfactory for most work, and they have the advantage that no sharpening is required. Figure 2-1 shows four types of knives.

OILSTONE

A fine-grained oilstone is necessary to keep knives sharp, other than the snap-off type. Sharpening on the stone is best followed by polishing with a mildly abrasive lapping compound. Frequent polishing will keep the knife sharp between sharpenings with the stone.

HOOK

A hook is convenient for holding the work during braiding. The hook should be placed in a wall or attached to some other firm support a few inches above elbow height.

Fig. 2-1. Many types of knives can be used for leather work, from an ordinary utility knife to a specialty knife from a leather supply store.

FID

A fid is needed for making knots and for splicing. A heavy awl with the point rounded off is satisfactory, or an ice pick will also do the job. If the handle is flat (or made flat) on one side it will prevent the fid from rolling when it is put down. Figure 2-2 shows three fid styles.

NEEDLES

Leatherworking needles are particularly useful for knot work, where weaving a strand of lace into a small opening made by a fid is both difficult and tedious. Needles of two types are available: a prong style, wherein a flat, split needle with two prongs holds the leather tip between the two halves of the needle; and a threading type, wherein the point of the leather strand is twisted into an internal thread in the hollow needle. Either type is satisfactory. In both cases, the leather strand should be brought to a gradual point where it is attached to the needle. It is helpful to thin the end of a strand made from medium to thick hide when using the prong-style needle.

ROLLING BOARD

A rolling board is used to roll round work. Rolling forms a smooth round surface. It will not eliminate irregularities or imperfections in the braid,

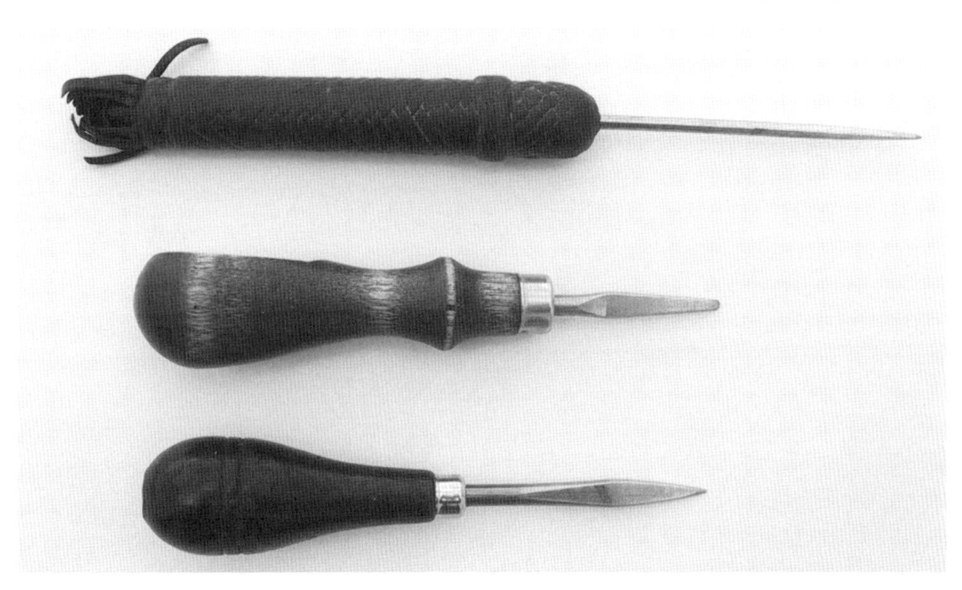

Fig. 2-2. Different items can be used as fids. Shown here, a common ice pick and two styles of fids available from leather supply sources.

but it may make them less obvious. A 2 × 4 is usually adequate, although a heavier section may be preferred for rolling large items. The board is wrapped with clean, strong, absorbent paper such as the heavyweight kraft paper used for grocery bags, and the work is rolled on a smooth firm surface that is also covered with paper. The paper absorbs grease from the surface of the work to help ensure that the braid rolls rather than slides.

CUTTING REST

A cutting rest is a help when cutting lace. The most convenient cutting rest for the braider using smaller skins such as kangaroo or calf is a board about 2 feet by 1 foot set just above elbow height. The skin is held in place by the forearm, and the lace being cut is held by the thumb and forefinger (fig. 2-3). A bar, such as a 2 × 4, about 5 feet long and set just above elbow height, is better for large skins or sides of leather. The Australian whipmakers, who cut the very large sides of red hide as well as kangaroo, generally prefer this. Finally, smaller skins can be held over the knees while the braider sits in a chair.

CUTTING GAUGE

For the amateur, a gauge helps in cutting lace evenly. Gauges are available commercially, or one can be made up. A simple but effective cutter may be made from a block of wood about 4 inches long, 2 inches wide,

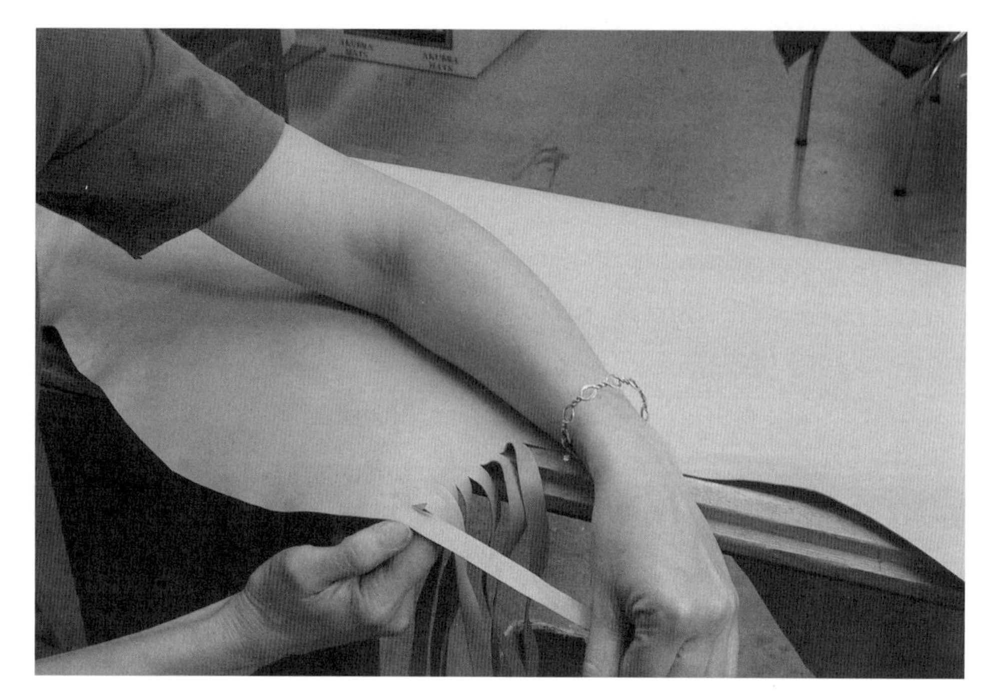

Fig. 2-3. Cutting rest

Adjustable gauge

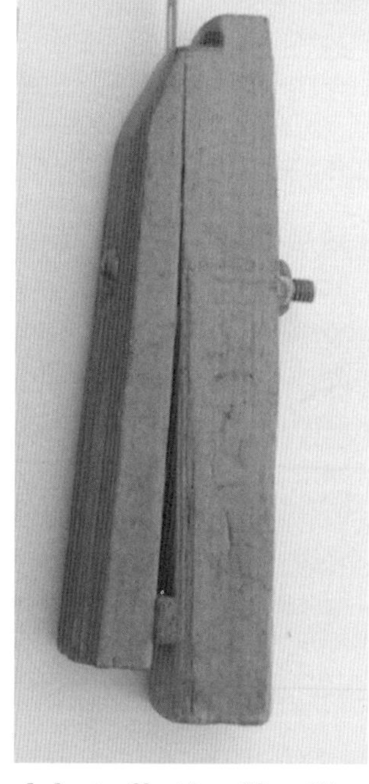

Rough but effective "fixed" gauge

Fig. 2-4. Cutting gauges

and ¾ of an inch thick. A ¼-inch-deep notch is cut into the top, extending ½ inch from the side. A slit is then made from the notch down the block to about ¾ of an inch from the bottom, or the side is cut off completely and wedged to slant it against the main block. A bolt is then put through a hole drilled across the width of the block about 1½ inches from the top. A wing nut on the bolt will then pull in the flap of wood to secure a single-edged razor blade held in the slit (fig. 2-4). The width of lace cut is determined by the distance between the notch and the blade. By extending the blade above the wood assembly, the gauge can also be used for the freehand cutting necessary to trim off scrap leather.

LEATHER SPLITTER

A leather splitter (fig. 2-5) is useful as a means of cutting lace to a uniform thickness, or of making thick lace thinner. The splitter is not essential for many kinds of braiding work, and its purchase may be deferred by most amateur braiders.

Fig. 2-5. Leather splitter

METAL STRAIGHTEDGE

A metal straightedge, such as a steel yardstick, is very helpful when cutting wider cores.

LEATHER SHEARS

Leather shears (or for light leathers, a suitable pair of scissors) are useful in cutting off ends of lace.

HOLE PUNCH

A hole punch (fig. 2-6) is useful when the braider wishes to put a yoke on the hook, make keyhole slits in leather, or do various types of lacing work associated with horse gear. It is not essential for most braided work.

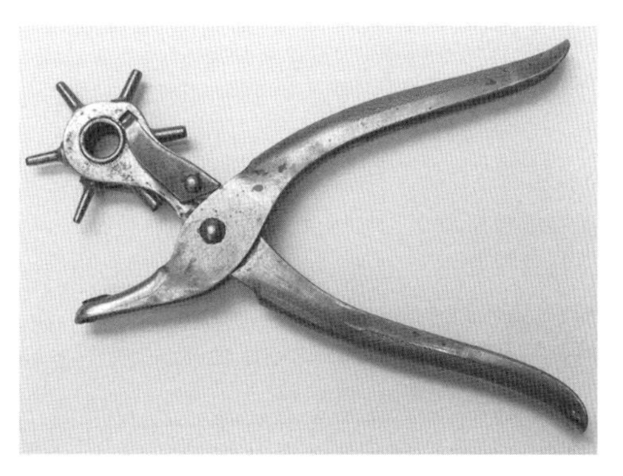

Fig. 2-6. Hole punch

PLIERS

Saddler's pliers or other suitable types are useful in pulling short strands, for example, when tightening the ends of the lace in knots (fig. 2-7).

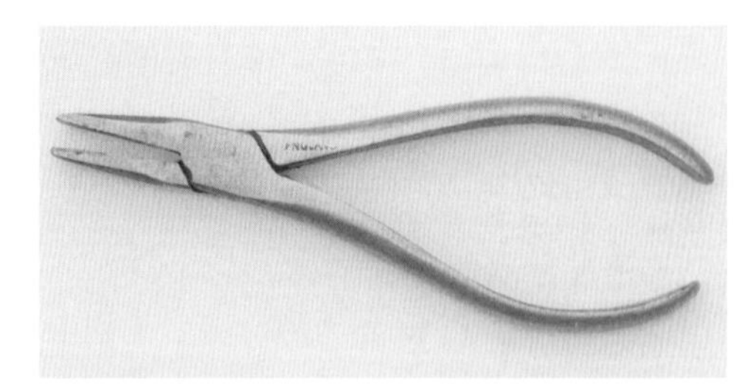

Fig. 2-7. Pliers

NOTEBOOK

The braider should keep a notebook with details of projects he or she has made. This record will enable the braider to repeat projects at a later date and will serve as an aid in determining needs for other projects as well. The width and length of lace used, the width or diameter and length of a finished article or parts of articles, the type of lace, the size and thickness of a core, the thickness of the leather, and the mode of paring are all details that can be noted. While the amateur should cut strands to a generous length to ensure that the project is not lost because of short strands, such generosity readily turns to waste, so the lengths cut should be kept as short as experience allows. The notebook allows the braider to accumulate the necessary experience efficiently.

The notebook also may be used to record details of braiding encountered elsewhere. A few dimensions taken from a well-designed set of reins and headstall, for example, would be invaluable to anyone attempting to make them for the first time.

CHAPTER THREE

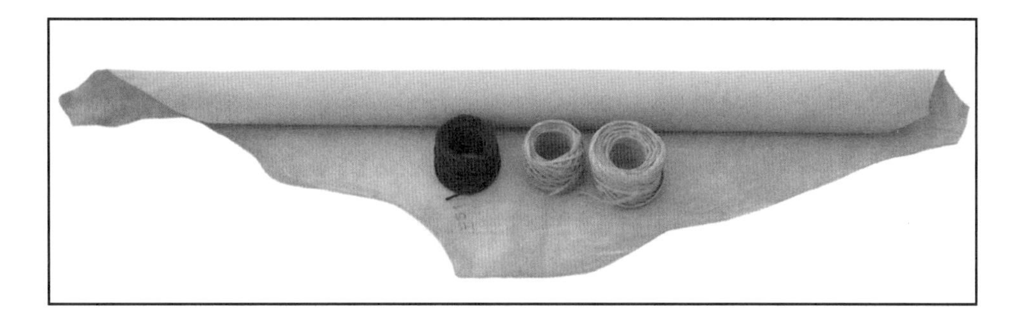

Materials

LEATHER

The leather used in braiding should be strong, supple, and capable of being stretched to a reasonable degree. For coarse braided work, latigo (or redhide as it is known in Australia) is generally satisfactory and can be obtained in a wide range of substance (thicknesses). For finer braided work kangaroo leather in a vegetable (bark) tan is clearly the best choice. Kangaroo has excellent strength and stretches well. It is a thin leather and cuts in a clean fashion. The projects in this book are all designed to be done in kangaroo leather. If kangaroo is not available, calf or kip (the skin from an older animal), again in a vegetable tan, may be substituted with somewhat inferior but still satisfactory results.

The leather should be well dressed, preferably drum stuffed at the tannery (grease is worked into the leather in a rotating drum), to ensure that the proper greases are fully worked into the leather. The leather dressing allows the leather fibers to bend and move when under stress, making the leather stronger and more supple.

BRAIDING SOAP

The use of braiding soap, an emulsion of fat in a soap-and-water solution, is necessary to allow the leather strands to slip into place during braiding. The water in the solution conditions the lace to allow greater stretch and to permit the finished braid to be rolled to a smoother surface. Following are directions for making braiding soap:

The ingredients of braiding soap are soap, fat, and water. A pure soap should be used, not a detergent or facial soap. Lard is satisfactory as the fat and is readily available in one-pound blocks. The proportions to use are 2 pounds of fat to 8 ounces of soap and 3 cups (24 fluid ounces) of water. A large coffee tin with a plastic top is convenient for making the braiding soap and keeping it, as the soap tends to dry out if left uncovered. The quantities indicated suit a 3-pound coffee tin.

Grate the soap and dissolve it in the water, bringing it almost to boiling to do so. Take care not to overheat the solution and cause it to boil over. Place the fat into the hot soap solution and heat until the fat is melted, again taking care not to allow it to boil over. Remove the mixture from the heat and use a food blender to mix and emulsify the fat. Allow the soap to cool, then mix it thoroughly once more. The finished soap should have a light, creamy consistency.

TALCUM POWDER

Talcum powder may be used along with braiding soap or by itself to lubricate lace for braiding. Used on the hands in conjunction with braiding soap on the strands, talcum allows much faster braiding by reducing the tendency of the strands to stick to the hands. It is often used alone for braiding hatbands, where grease may be objectionable in the finished article.

SHELLAC

Shellac—orange by tradition, although white provides an equal sheen—is used to finish braided work. Shellac does not completely seal the surface, permitting subsequent applications of leather dressing to go into the leather. In some work, such as braided coverings of stiff handles where the leather is not flexed and minimum replenishment of the oils in the leather is required, a leather lacquer may be used. Shellac and shellac thinner are available at paint stores.

HARDWARE

Various types of hardware are needed—swivels for lanyards, bolt snaps for dog leads, buckles and dees (D-shaped rings) for belts, and rings for bridles. Care should be taken to use only good quality hardware and to ensure that the hardware used in any item fits properly into the overall design.

WAXED THREAD

A heavy waxed thread is used to ensure that the string forming the base for knots or used to whip the ends of a braid holds well. This type of thread may be purchased, or any light string may be treated with beeswax or a similar product such as surfboard wax.

LEATHER DRESSING

A good quality leather dressing is used both to maintain braided items and to grease craft lace before braiding it. Dressings containing animal fats (which become rancid) or light oils (that make the leather stretch badly) should be avoided. For long-term protection, as for items held in a collection, a dressing that prevents attack from the acids in the air and also inhibits oxidation of the leather is essential. Pecard Leather Dressing is one such product.

CHAPTER FOUR

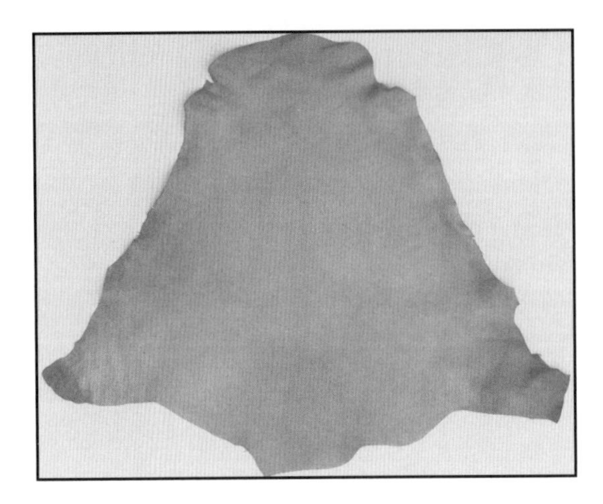

Leather

GENERAL CHARACTERISTICS

The characteristics of an animal's skin vary from part to part to suit specific needs. Leather made from the skin varies accordingly. The braider, in order to produce good work economically, must learn to cut the lace for braiding so as to make the best use of these variations. Since kangaroo hides or calfskins are relatively small, any long lace must be cut in a spiral fashion from a uniform portion of the skin. Shorter lengths are usually cut from the less-uniform portions. The experienced craftsman is able to cut a skin into lace for a variety of projects, with the lace for each project well suited to its use and with minimum waste. The beginner should form a plan for cutting the entire skin before starting.

Weak leather is usually found in those places where the skin of the animal needs to stretch readily, as at the neck, around the belly, and at the flanks, where the skin stretches to allow the legs to move. Firm, strong leather is normally found on the back, where the skin is not required to move a great deal. The leather from the neck and legs is moderately strong along the length of the neck or legs, but weak and stretchy crossways.

KANGAROO SKINS

Kangaroo is by far the preferred leather for fine braiding. Kangaroo offers several advantages over calf or kip. It is much stronger, and it cuts with cleaner edges. It shows scars on the surface, so those areas can be avoided when cutting the lace. Calfskin heals over and hides scars, so the lace may have weak spots from such hidden scars. Kangaroo lace maintains the shape of its cross section when stretched. In calf lace, the edges tend to curl up around the grain side when the lace is stretched.

Kangaroo skins are roughly triangular in shape, with heavy back legs and comparatively small front legs. The leather from the flanks (those parts of the skin that stretch when the legs are moved) is thin and weak, but the leather from the legs is reasonably strong along the length of the leg. The skin on the belly is weak and often scarred, and the area near the neck tends to stretch heavily. The strongest and most uniform section is the lower central portion of the back. Figure 4-1 shows the characteristics of kangaroo leather taken from different parts of the skin.

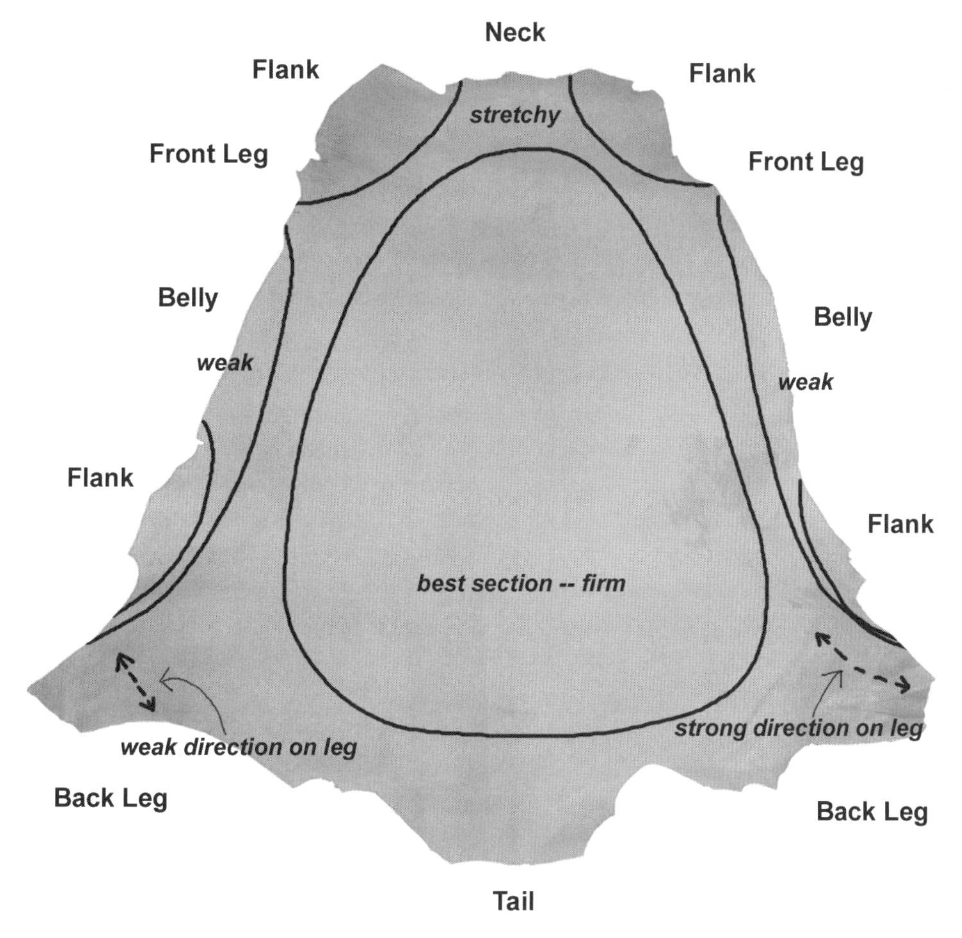

Fig. 4-1. Kangaroo skin

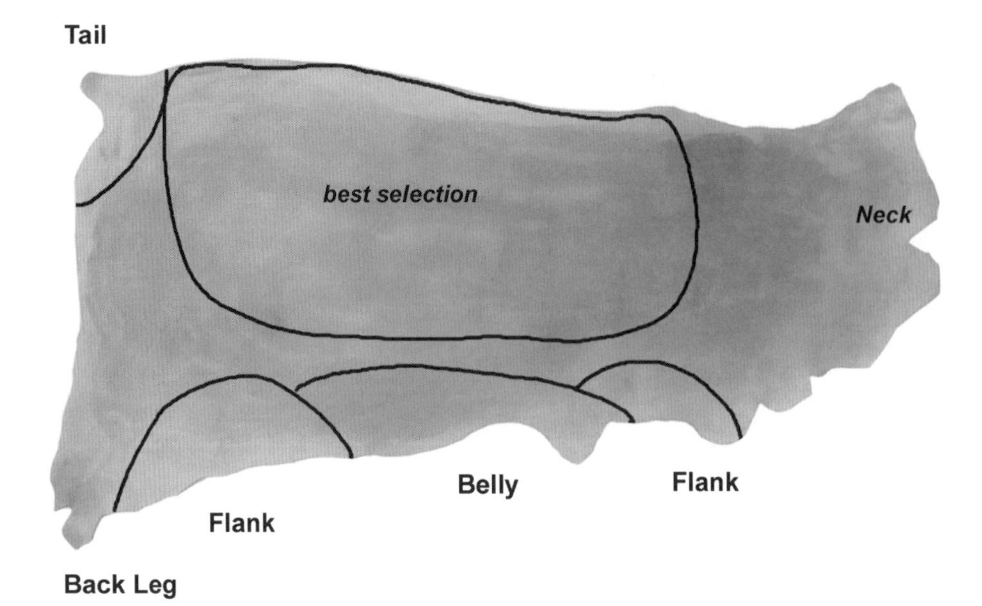

Fig. 4-2. Kip side

In cutting kangaroo, the weak flank and belly leather is first trimmed off, either as waste or to be used as part of the filler for inner cores. Then lace is cut for projects that use short lengths but do not require maximum strength or uniformity. This lace is usually cut by starting at a back leg and proceeding toward and around the neck. The objectives in cutting this lace are both to trim off the weaker and more stretchy parts and to leave the central section not only uniform but also well shaped—more circular than angular—so that long lace may be cut from it economically, without the wastage developed in cutting around sharp curves. The more stretchy lace trimmed off the outside may be cut somewhat wider than needed, and after stretching, it may be pared to the uniform width desired.

CALFSKINS OR KIP SIDES

Calfskins are very small and usually leave a lot of waste when cut. Kip, the skin from a somewhat older animal, is generally preferable. Kip is available in sides (fig. 4-2). The poorer leather found near the tail, on the back leg and flank, on the belly, on the front leg and flank, and on the entire neck must be removed first. The best leather is found from the tail end to the shoulder, from the top of the back to about half way down to the belly. Where possible, lace should be cut from the length of the side. To get more length, cutting should start at the back leg and proceed past the tail, along the back to the shoulder, and onto the neck, if the strength there is sufficient. Very long strands can be cut by going around the best section of the back.

PRECUT LACE

Lace may be purchased already cut. Precut lace is a convenience when only a small project is to be undertaken or when the beginning braider wishes to try braiding before learning the techniques of cutting lace. See figure 4-3. Hand-cut kangaroo lace is preferred. It is cut around the hide; flanks, scar tissue, and weak spots are removed during cutting. However, the lace may still vary in its properties and may have occasional flaws or weak spots. Lace cut by the use of jigs or machines is less expensive. It is usually cut from circles; it tends to be less uniform in thickness and contains more scars and flaws, since these cannot be removed during cutting. This lace is often made into continuous lengths by gluing together shorter lengths. When precut lace is used for projects, the lace should first be stretched to break any weak or glued parts, then sorted into firm and less-firm lengths, even and less-even, and so on, so that the lace characteristics can be matched to the projects. The quality of machine-cut lace varies significantly, depending upon the quality of the skins used and the amount of trimming done prior to cutting. The better quality lace has fewer flaws and joins, and it will be more uniform in its characteristics.

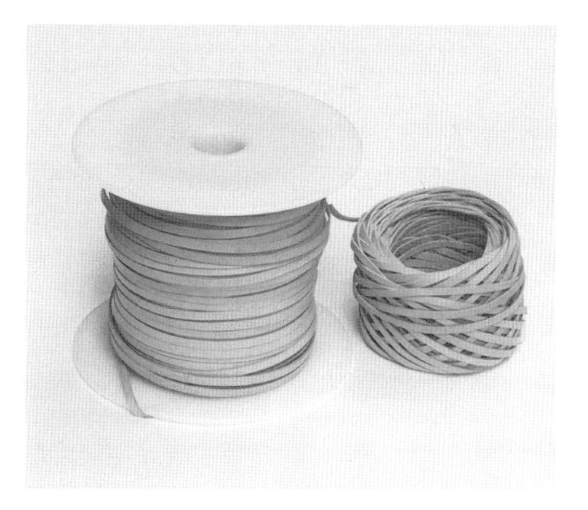

Fig. 4-3. Precut lace (machine-cut or hand-cut) is available in a range of leather types, widths, and grades.

For braiding, lace from a drum-stuffed skin is best; the grease improves the strength and braiding characteristics. Craft lace, cut for more general craft work, has less grease in it, so it can be dyed more readily. Craft lace should be greased well with leather dressing before being used for braiding.

CHAPTER FIVE

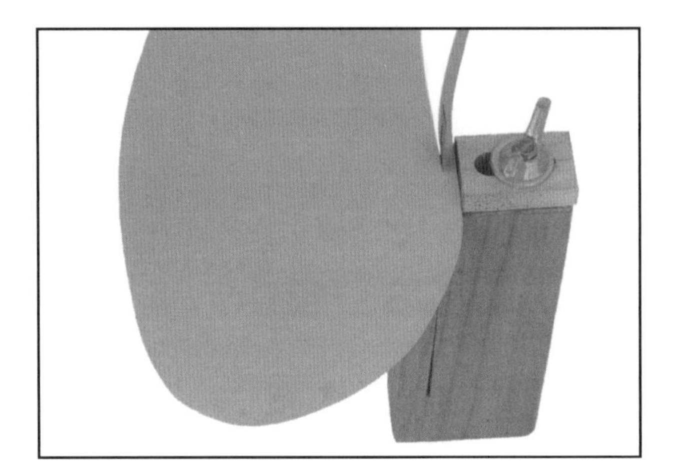

Cutting Lace

OVERVIEW

In an apprenticeship, a beginner would first learn to braid lace that had been cut and pared by a trained craftsman. Beginners working on their own must of necessity either use precut lace or they must cut and prepare the lace before braiding, using a bootstrap approach of learning to cut, pare, and braid all at the same time. Most braiding projects, other than whips or other tapered items, can be done using precut lace. For advanced work there are advantages to cutting lace not only from selected skins but also specifically for a project. Those who prefer to use precut lace should read this chapter closely to gain some understanding of the variations inherent in lace cut from a kangaroo skin; this preparation will enable braiders to make better use of the lace and also to avoid potential problems.

Kangaroo skins can be held in the lap or placed on a table or bar for cutting. Large skins are best handled on a horizontal bar set just above elbow height.

Before lace is cut from a skin, the rough edges and unusable flank portions must be trimmed off with a sharp knife. As the lace is being cut, scars or other undesirable flaws in the skin should be removed. The lace should be cut past a scar but not cut into it. A smooth cut should start before the scar and curve smoothly back to the edge of the skin to remove the scar and leave the skin so the next lace can be cut without abrupt curves. See figure 5-1. A gauge with a razor blade can be used to cut light

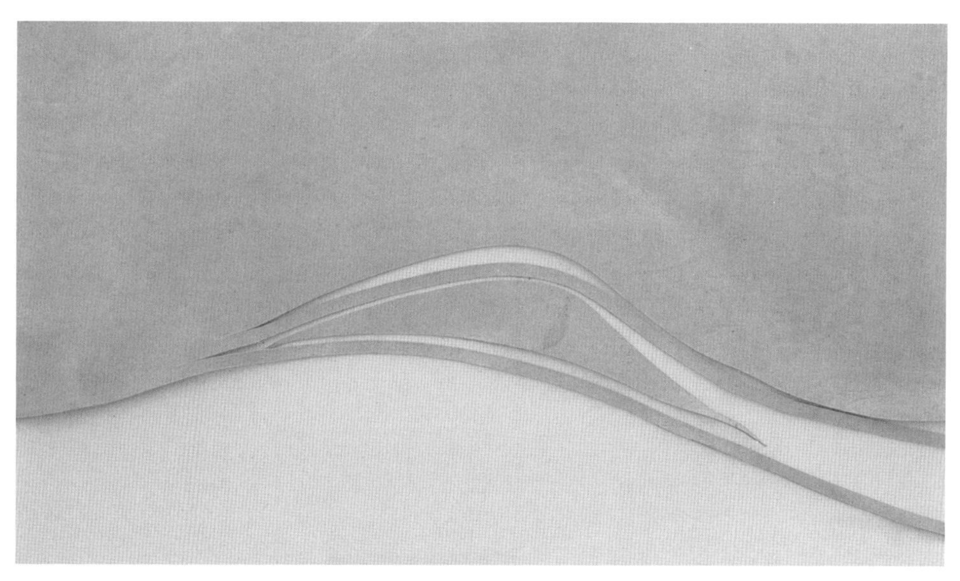

Fig. 5-1. Removing scars during lace-cutting

skins. The razor blade is moved up or down in the gauge as needed to present a fresh sharp cutting edge, and it is replaced as necessary. A sharp razor will normally start a fresh cut readily if it is drawn down a little as it is pushed against the skin. Alternatively, a start can be made by cutting the skin against a board with a knife. Once started, the lace is held between the thumb and index finger of the left hand (fig. 5-2). The skin is supported on the bar or table and held in place there by pressure of the left forearm, with the edge being cut extended beyond the side of the bar or table. The gauge is pulled along the skin with the right hand. Those who are left-handed may wish to adapt these directions accordingly.

The essential factor in cutting lace is that the lace must be kept in tension, pulled directly opposite to the direction of the knife blade cutting the leather around the curve, i.e., tangential to the curve of the edge of the skin at the point of cutting. If the tension is dropped the lace will be cut too narrow or may be cut off completely. If the direction is not correct the lace will be cut too narrow or will bind in the gauge.

Good lace cannot be cut around a sharp corner or a very tight curve. The tightness of the curve around which lace can be cut successfully depends upon the width of the lace being cut, the stretch in the lace, and the quality of lace desired. A little experimentation will help the beginner gain an understanding of this factor. A tight curve will affect the way the lace behaves during paring, so strands should be cut from corners of different curvature, then stretched and pared to establish the

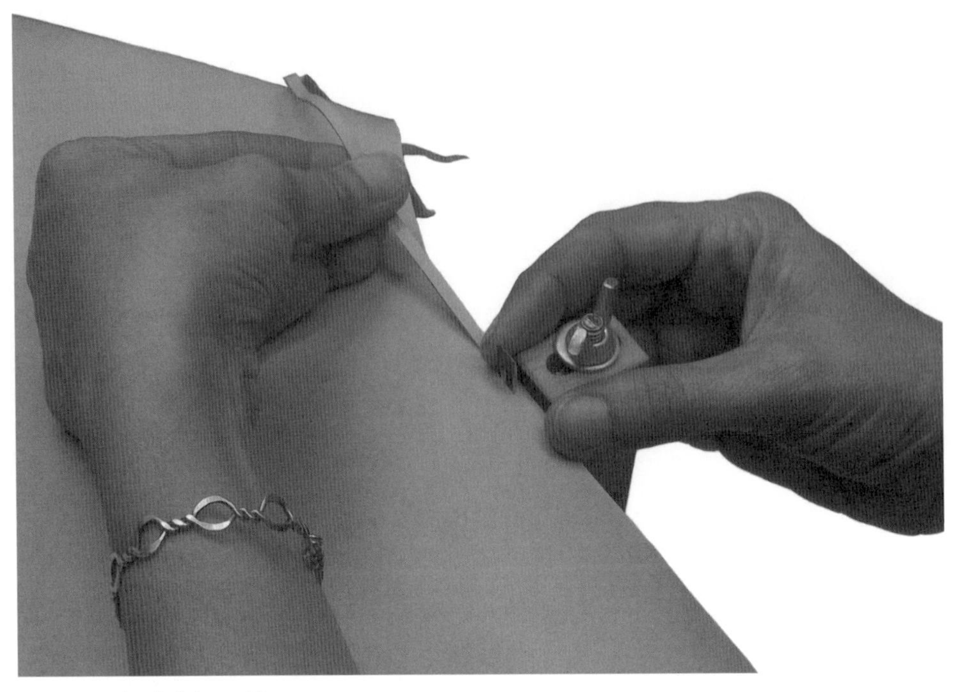

a. straight section

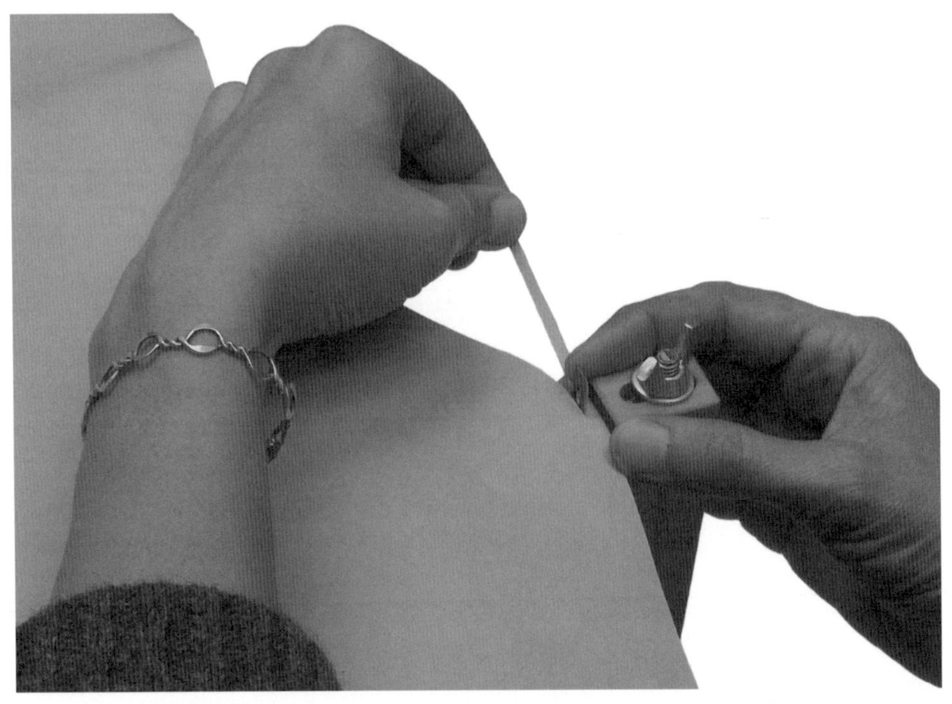

b. curved section (note the angle at which the lace is pulled)

Fig. 5-2. Cutting with a gauge

maximum curvature that can be handled. As a general rule the maximum curvature recommended when cutting ⅛-inch lace is about the curvature of a 2-inch circle. For ¼-inch lace, curves no tighter than those of a 4-inch circle are suggested. The effects of curvature show up more on flat braiding than on round braiding.

Another way to cut lace is to use the thumbnail or the thumb itself in conjunction with a knife (figures 5-3 and 5-4). This method has several advantages over the use of a set gauge. The width of the lace may be varied readily to make tapered strands or to compensate for stretchy parts of the skin; the strands may be cut with beveled edges, and imperfections or variations in the skin are more readily sensed by the cutting hand. The disadvantages of cutting with the knife and thumbnail are that it requires experience to maintain a consistent width or to cut a predetermined width accurately. However, the technique is by no means beyond the amateur braider. Accuracy in setting a predetermined width on a gauge is often negated by the variation in skin thickness or stretch, so the advantages of using a thumbnail far outweigh the disadvantages. The amateur should practice with both gauge and thumbnail in order to develop use of the latter at an early stage.

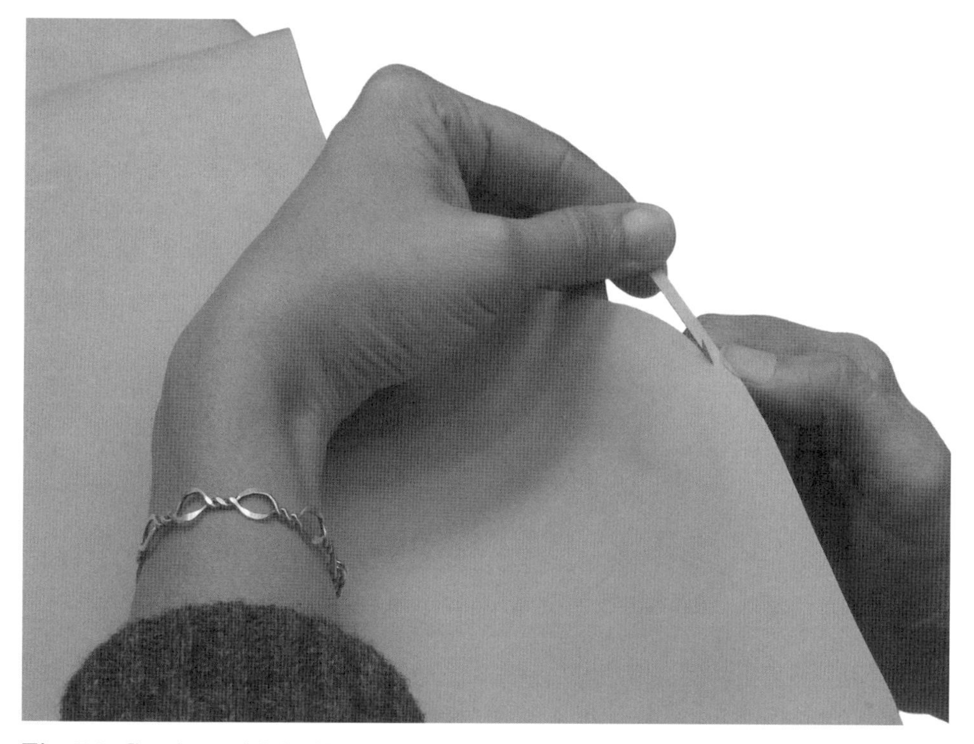

Fig. 5-3. Cutting with knife and thumbnail

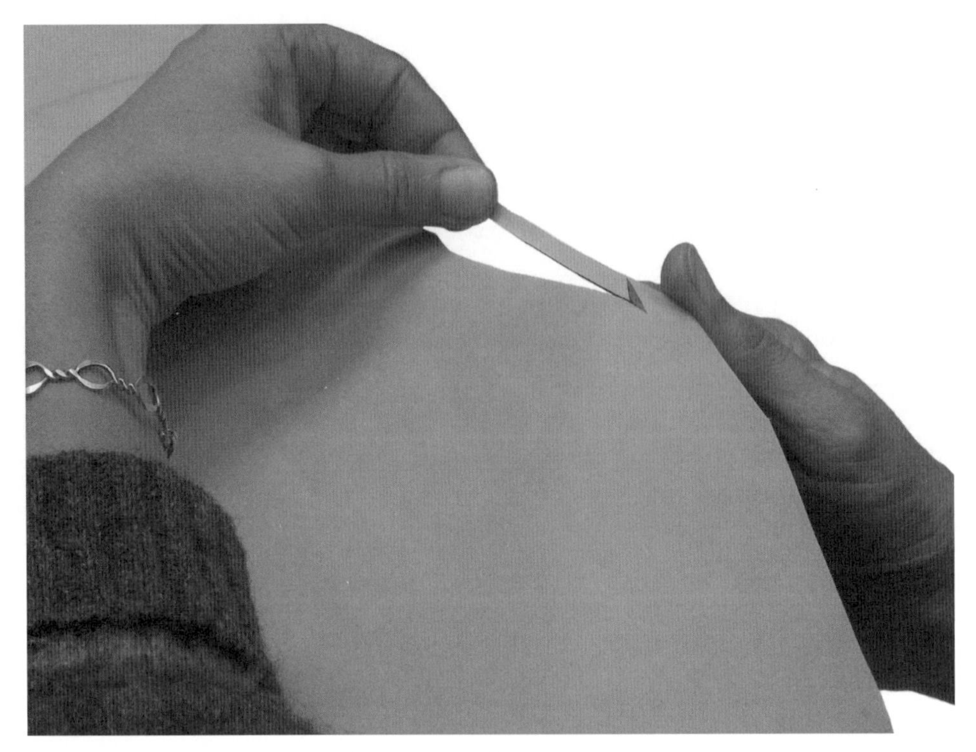

Fig. 5-4. Cutting with a knife and thumb

In cutting kangaroo skin, the edge of the skin usually rides under the thumbnail. If the nail is long and the lace is not too wide, the knife blade may rest against the thumbnail. By variation in the angle and position of the knife as it rests against the nail, a wide range in lace width may be cut. With a very short thumbnail or when cutting heavy leather or very wide lace, the leather rides against the thumb and so is less firmly positioned. The forefinger rides against the underside of the skin in both cases. In cutting, the lace is held taut at an angle directly opposite the direction of the cut and tangential to the curve, just as when using a gauge. Changes in the width of the lace or the firmness of the skin are readily sensed by the thumb. Imperfections on the underside of the skin may be felt by the forefinger.

GETTING STARTED

The objective of this exercise is to learn the characteristics of the lace that is cut at each stage or step in the process of cutting up a skin. The beginning braider who is not familiar with paring or braiding should refer to chapters 6 and 7 to learn how to make a four-strand round braid, so that he or she will be able to evaluate the lace cut from various parts of the skin. The lace should be cut ⅛ inch wide until all the braids showing the characteristics of the various sections have been made. In learning to cut lace, the first skin should be cut as a planned experiment. At each stage the lace cut should be stretched, pared, braided, and evaluated. Notes should be taken and the braided samples retained for comparison with other samples. Figure 5-5 shows the shape of the skin with the edges stretched.

Cutting the Unusable Weak and Stretchy Leather

First the edges of the skin should be checked, stretching short sections at a time between the hands. This will show up the weaker, more stretchy parts, which should be trimmed off along with any rough or ragged edges of the skin, starting at one back leg and moving around the neck to the other back leg. This initial trimming should be minimal (fig. 5-6). The skin should be checked again, stretching short sections between the hands to show up the very stretchy parts of flanks and other weak places. The worst of these must be cut out. This process provides

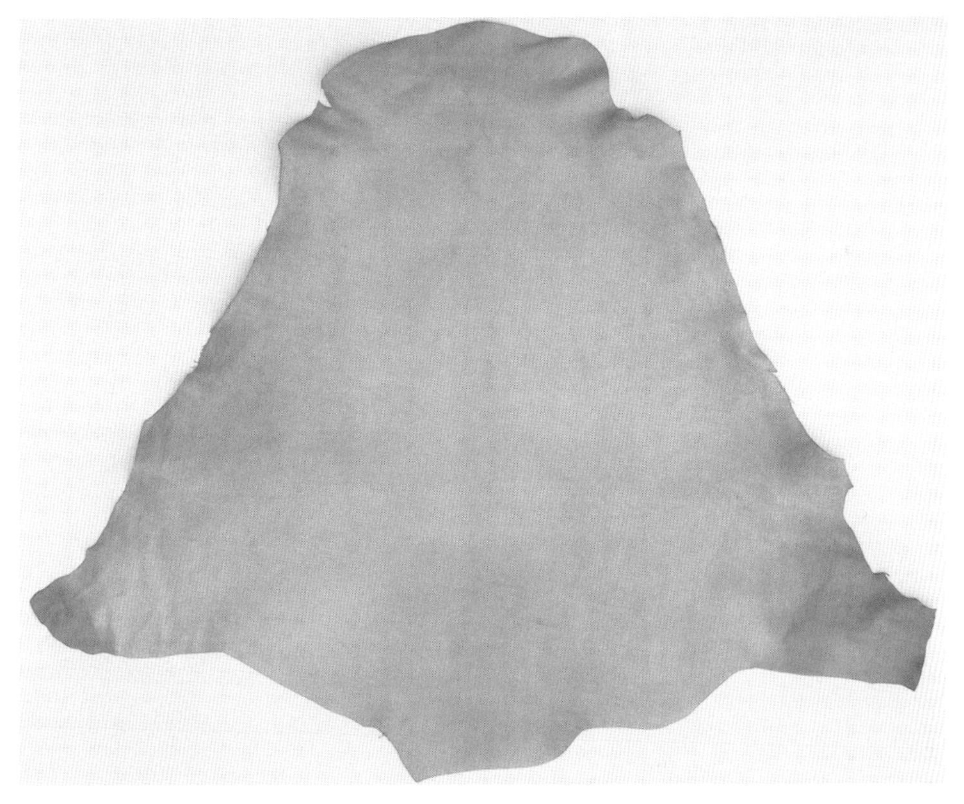

Fig. 5-5. Skin with edges stretched

an opportunity to learn to cut lace with a knife or gauge. Weak sections are trimmed out by cutting a series of laces.

Next a strand should be cut, starting from one back leg and going around the neck to the other back leg. The strand should be tied onto a hook and slowly stretched so that the weaker (more stretchy) parts can be found. The corresponding parts of the skin can then be stretched again between the hands or over a corner of a bench to minimize the stretch. The amount of curvature of the corners and the effect of curvature on the lace should be checked; curves that are too tight should be rounded more. Curvature should be watched throughout the cutting process. Another lace can then be cut from the prestretched skin, and it can be stretched to observe if any improvement has been gained. Excessively weak or stretchy leather may show up, indicating places where more flank or belly leather should be cut out. These first strands can be used to learn or practice paring.

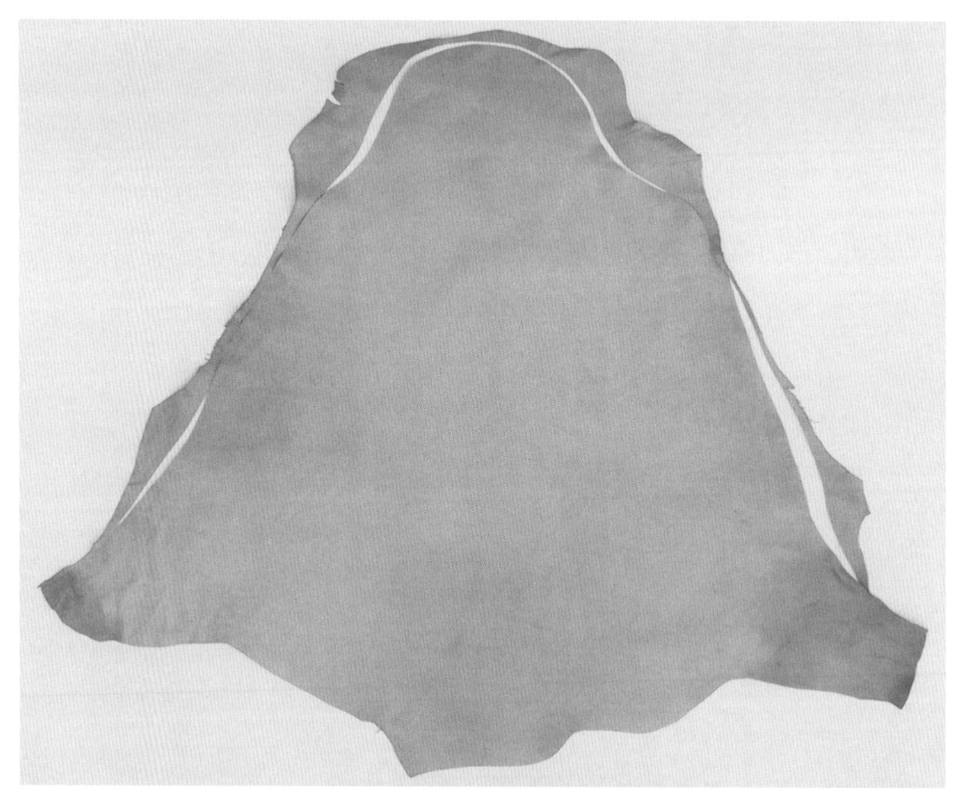

Fig. 5-6. Initial trimming to remove the weaker, more stretchy parts of the skin

Cutting Lace for Test Braids

Although the skin is probably not fully trimmed at this point, two test braids can be made. The first will demonstrate the effect on the braided product of the differences in the skin at the flanks, belly, and neck. The second test braid will demonstrate how the effect of these differences can be reduced by repositioning the lace.

For the first of the test braids, a set of four strands should be cut, starting at the back leg and going around the neck to just beyond the opposite front leg. This set should terminate in a yoke, that is, all the strands should remain attached to a solid end strap (fig. 5-7). To make the end yoke the first three strands should not be cut off, but rather left in line with each other, hanging on the skin. The fourth strand should be continued past the place where the first three strands stop and out to the edge of the skin to form a triangular yoke (fig. 5-8). The yoke is then tied to the hook and the strands are stretched, taking note of differences in stretch characteristics along the strands and between the first strand cut and the last. The strands are pared; braiding soap is applied, and the

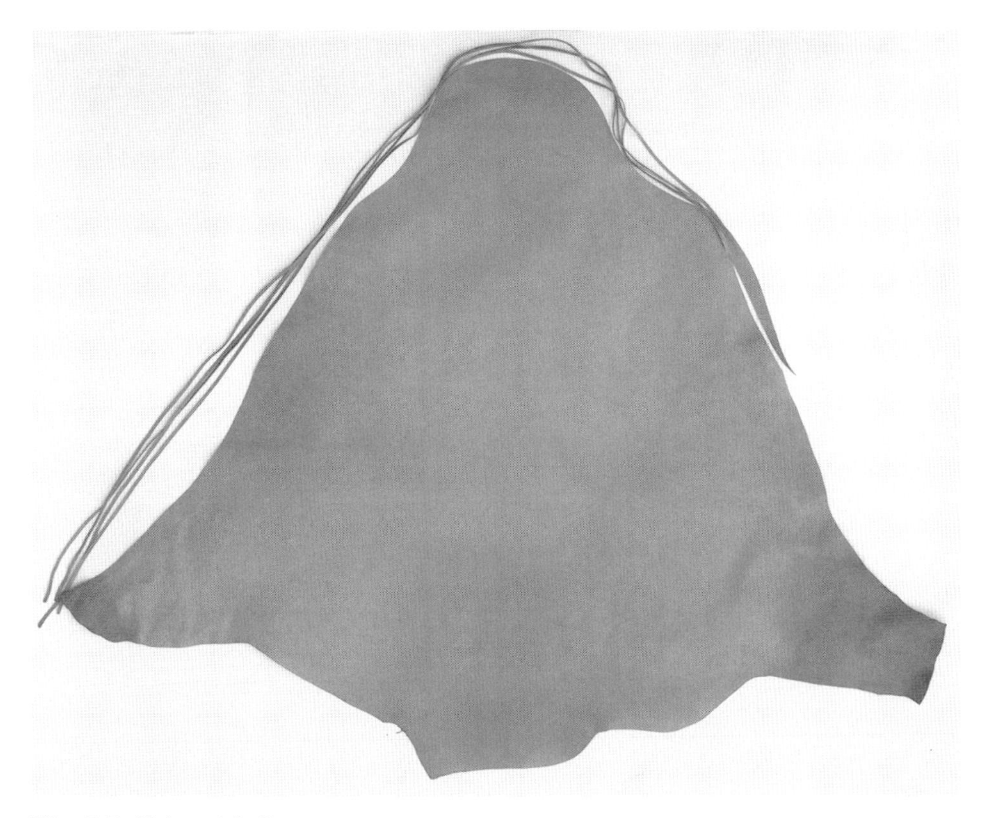

Fig. 5-7. Skin with first set cut

four strands are braided into a round braid, which is then rolled. Changes in diameter and variations in the appearance of the braid should be noted, and the piece saved as a sample for later comparison with others.

a. Stop the first three strands where the yoke will start.

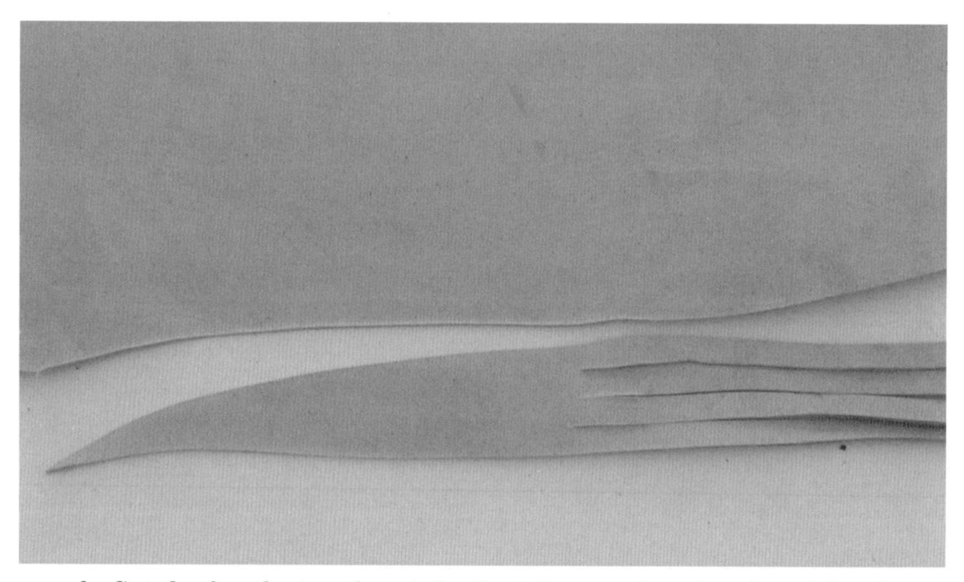

b. Cut the fourth strand past the first three and to the edge of the skin to form the yoke.

Fig. 5-8. Cutting a four-strand set finishing at a yoke

Next a second set of four strands should be cut symmetrically with the first set, starting with a triangular yoke just before the front leg, extending over the neck to the other back leg (fig. 5-9). To make the yoke for this set first make four slits where the strands are to start. Cut the first three strands for a short distance and then cut the triangular yoke, starting at the edge of the skin, up to the slit marking the start of the fourth strand. The fourth strand can then be continued to bring it in line with the first three, and then all strands cut in sequence (fig. 5-10). The yoke is placed on the hook, and the strands stretched and pared. This time two of the strands should be cut off the yoke and the opposite ends of these tied to the hook. The braid will be made with two reversed strands, so that the flank sections of two strands will be placed at a different position in the braid than those of the other two strands. Braiding soap should be applied and the four strands braided into a round braid, crossing the two strands that are attached to the yoke so that the two yoke strands will be in opposing spirals. The finished braid can be rolled and its appearance compared with the first sample. This braid will not vary in thickness nearly as much as the previous braid (where the four strands remained in

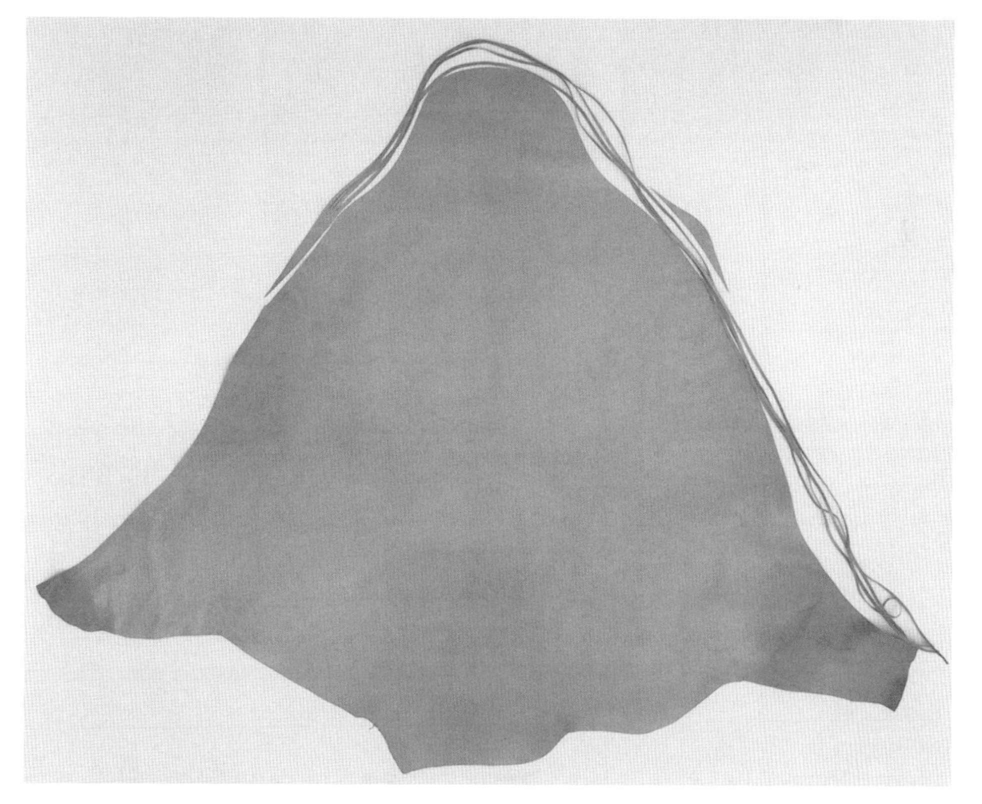

Fig. 5-9. Skin with second set of four strands cut

a. Cut four slits in the skin where the strands will start.

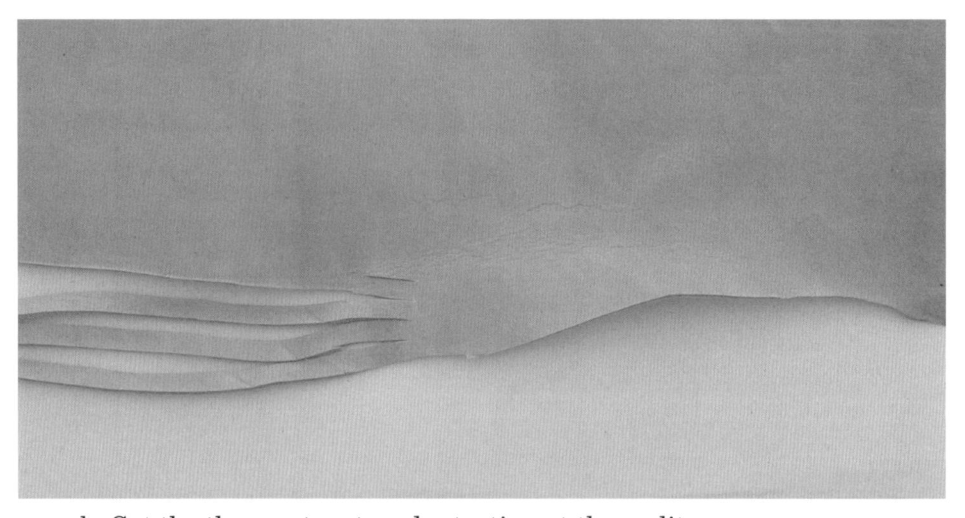

b. Cut the three outer strands starting at these slits.

Fig. 5-10. Cutting a four-strand set starting at a yoke

the same relative position as the flank or belly leather from which they were cut). Reversing half the strands has an averaging effect. This braid should be saved for later comparison with others.

The same procedure can be employed to evaluate the tail section of the skin. In general this leather is much firmer than the belly or neck leather, but it does have a stretchy flank adjacent to the tail, and stretch in the tail itself differs along its length. The edges of the tail section should be trimmed from one back leg to the other (fig. 5-11). Then a set of four strands may be cut from one back leg to the other (fig. 5-12). Except for reversing the direction of two strands as was done with the pre-

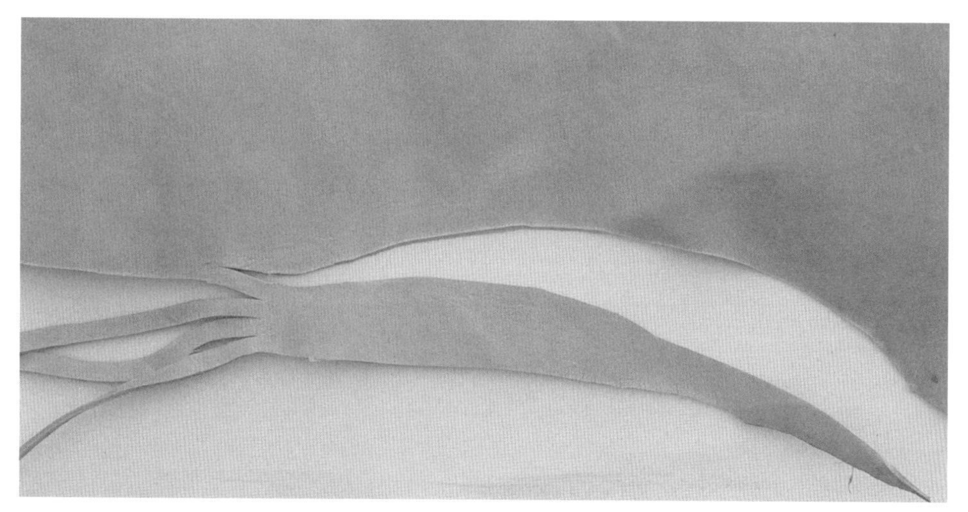

c. Cut the yoke up to the fourth slit.

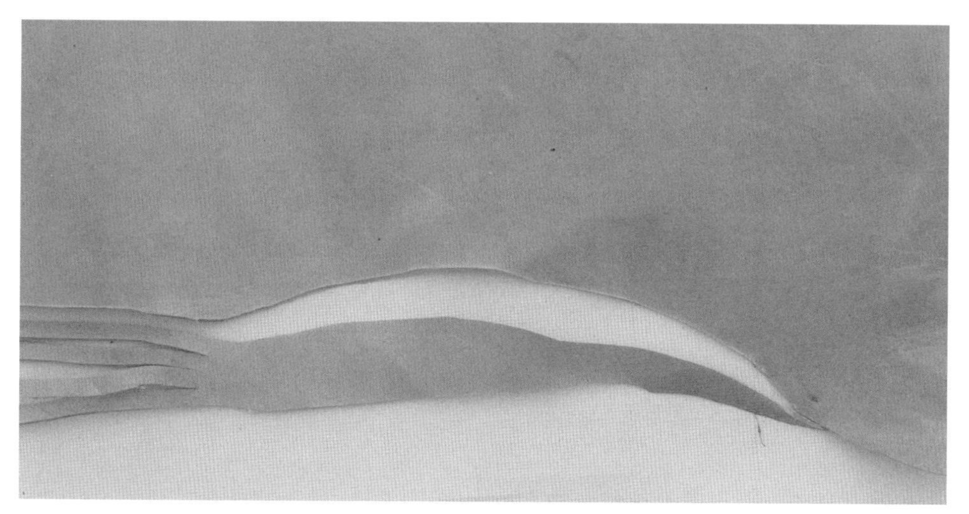

d. Continue cutting the fourth strand.

vious test braid, the process is the same: the strands should be stretched and pared, braiding soap applied, and the braid rolled after completion. This sample should be compared with earlier samples. This braid will be thicker than the others in the central section where the strands were cut from thicker leather. Once the weak parts next to the tail have been removed, this end of the skin is all prime leather.

Cutting Lace for Projects

At this stage there are probably still significant differences between the lace cut around the neck and that cut around the tail; there may still be

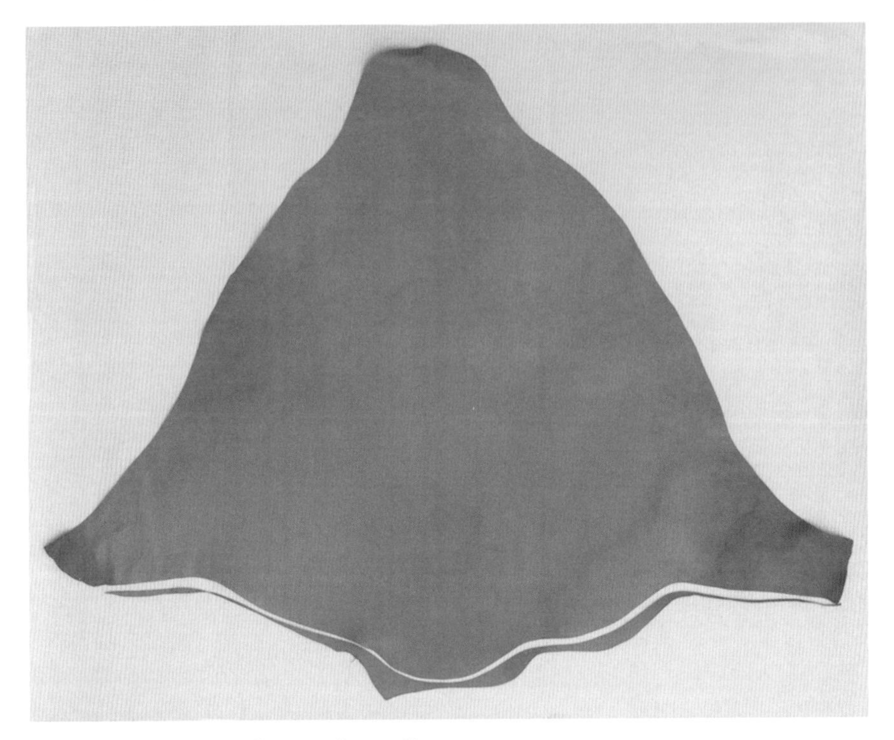

Fig. 5-11. Skin trimmed over the tail

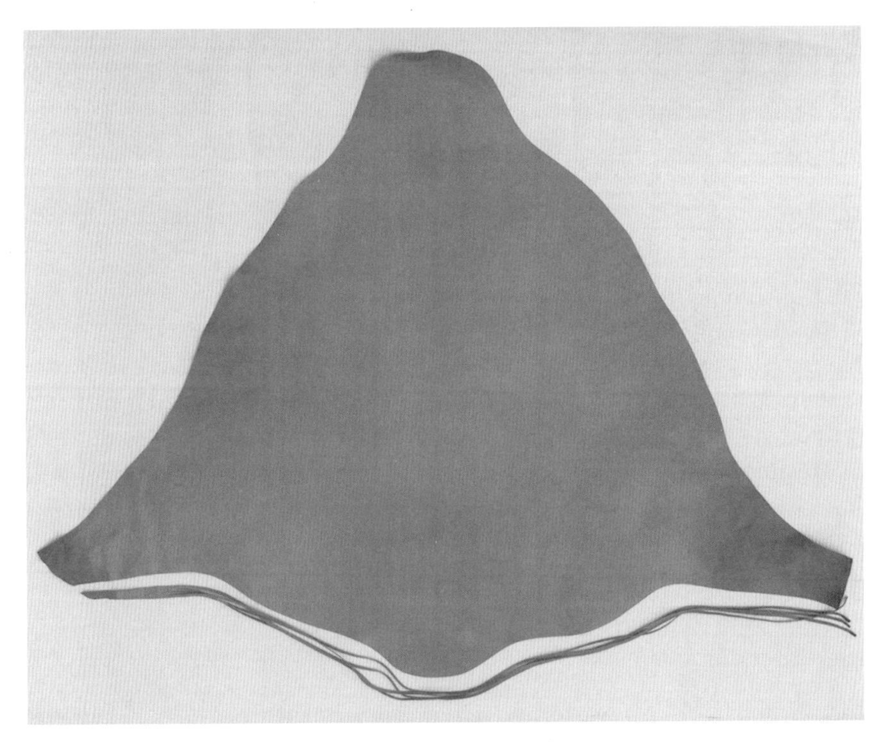

Fig. 5-12. A set of strands cut along the tail section

some very stretchy flank leather that needs to be removed, and portions of the back legs may not have been cut into lace. In order to even out these differences and use up the leg leather, more lengths of lace can be cut, starting from one back leg and going around the neck and then to the other back leg (fig. 5-13). Lace cut at this stage is usually suitable for less demanding round work and could be used in the novice braider's first projects. When the leg leather is used up, the leather in the neck end and the tail end most likely has sufficiently similar characteristics so that long lace could be cut.

Cutting Lace from the Inner Uniform Leather

The skin should be rounded off at the back legs. Short sections of lace can be cut around the area where the leg has been removed to check that sufficient material has been cut off so the lace will be uniform. One strand should be cut around the entire skin and stretched to see if any parts show excessive stretching. If not, the strand should be pared to make a uniform lace, cut into four pieces, and braided. This sample can

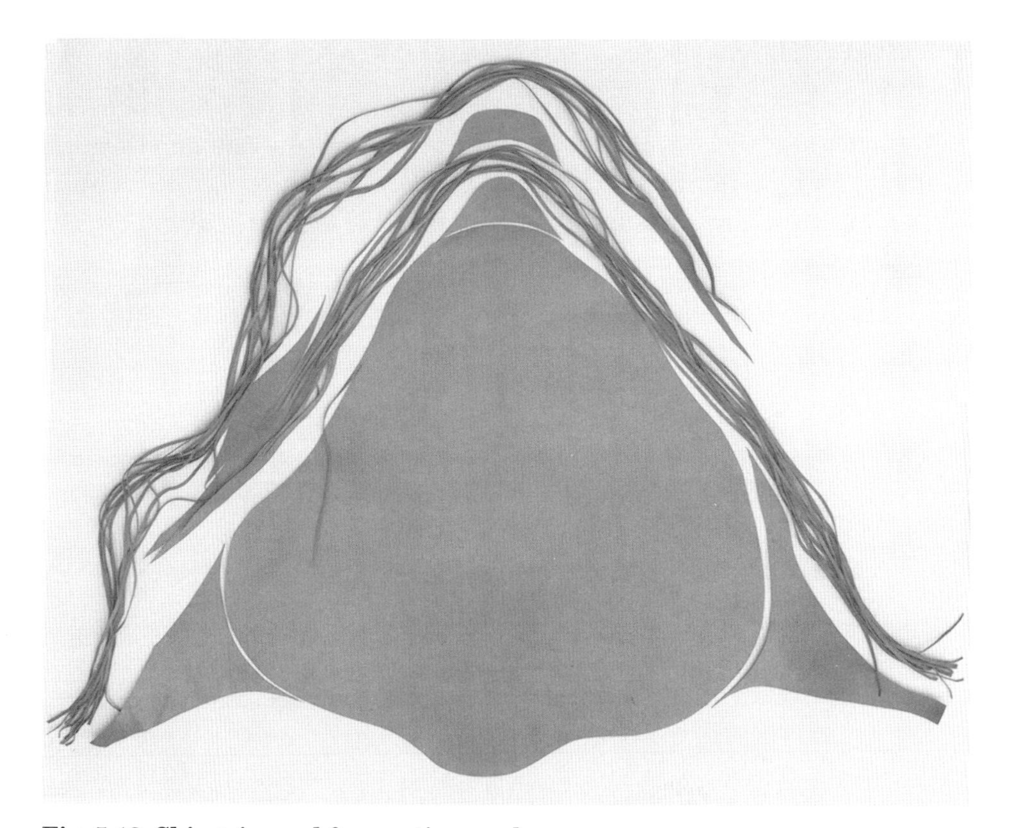

Fig. 5-13. Skin trimmed for continuous lace

be compared with earlier samples. If all looks good, the process can continue, cutting one long lace around the skin. The curvature around the protruding lobes should be watched, and the outer part of the lobes trimmed off as needed to avoid overly sharp curves. Throughout the cutting it is useful to test the lace cut around curves—pulling the lace to straighten it will show how tight a curve may be cut from each part of the skin.

Cutting on the Hook

When the skin has become too small to cut conveniently on a bar or table, it may be cut on the hook. The lace is looped over the hook and the free end is held by the left thumb and forefinger. The skin is supported very lightly by the other fingers of the left hand so that it may move as lace is cut. The knife or blade is kept in line with the hook as it cuts (fig. 5-14). The technique of cutting on the hook is readily acquired.

Carried out in this fashion, cutting the first skin is neither a daunting task nor a wasteful exercise. The first few feet of lace are cut from leather that is mostly unusable but the exercise serves as practice for cutting and paring, and it demonstrates differences in the leather. By cutting and braiding directly the greatest correlation can be made between the leather strands taken from different parts of the skin, their handling characteristics during the entire process from cutting through braiding, and the quality of the final braid. Most of the lace cut will be usable for future projects, and the final strand of lace cut around the entire skin should be usable for any type of work.

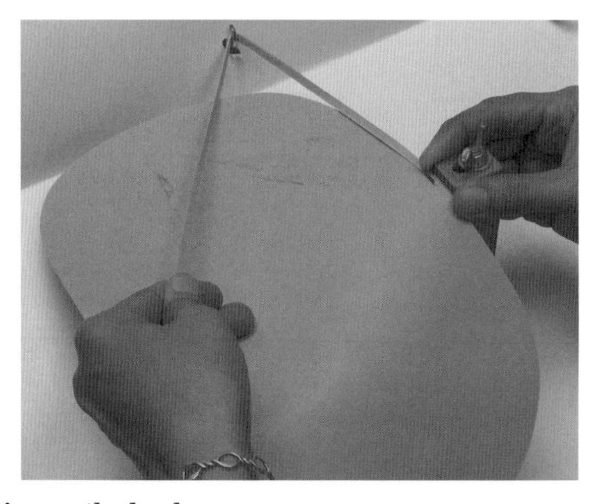

Fig. 5-14. Cutting on the hook

CHAPTER SIX

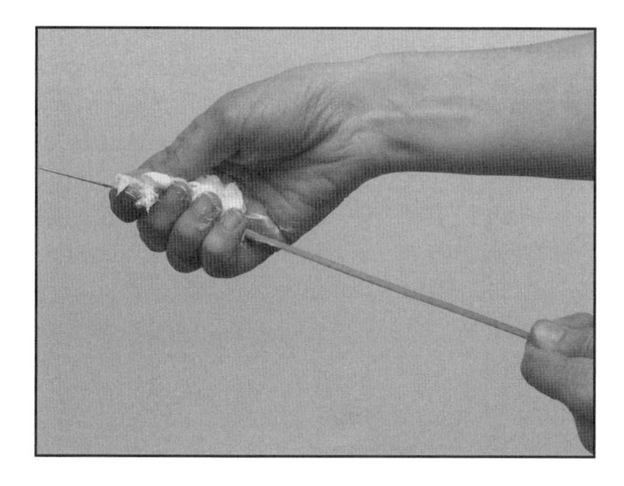

Preparing the Lace

OVERVIEW

Proper preparation of the lace to be used in a braiding project is a major factor in the ease with which the project is made as well as the achievement of a successful result. Lace that is too short or that breaks during a project is an obvious problem. Less obvious but very important is how well the lace will lay in place in the final braid. This characteristic is affected by the stretch and firmness of the leather, by the way the leather is pared, and by the use of braiding soap.

When lace has been selected or cut, it must be tested for weak spots, cut to length (if not already cut from the skin to the length needed), and usually pared. In testing for weak spots, the lace is tied to a hook and pulled, the pulling force being somewhat greater than the force to be used in braiding. This testing has two objectives. First, weak spots are found before braiding is started. It is much better to break a strand before braiding than during braiding. Second, the lace is stretched so that it will braid more uniformly, and the finished braid will retain its uniformity during use. Lace, particularly that cut from the outer edge of a hide, stretches more in some sections than in others. It is better to take out some of the stretch before rather than during braiding. Furthermore, if the lace is stretched before it is pared, the paring can better even out the width variation caused by stretching.

Paring, in which two corners of the lace are cut off, is customarily done for most projects. First, it allows the strands to lie closer to one another in the finished article and so produces a smoother finish. For a smooth finish on flat work and for round work in four strands, the lace is pared on both corners on the flesh side, leaving the grain side of the lace intact. This paring changes the cross-sectional shape of the strand from that of a rectangle with square sides to that of an isosceles trapezoid, with the sides sloping in from the grain side of the lace to the flesh side. For round work in six or more strands, two opposite corners are pared, leaving the cross section in the shape of a parallelogram. See figure 6-1. When braided, the strands overlap the adjacent strands partially where the sides are pared. For some products, such as belts, a rougher or more pronounced texture may be preferred, and for this reason the strands may be left as they are cut, not pared.

The second function of paring is to make the lace more uniform. Lace as it is cut around a skin varies in thickness and the amount of stretch as well as in actual width. When stretched, further variation is produced when thin or stretchy parts expand, or those parts cut around a tight curve straighten. Paring evens out the width of the lace and reduces the effect of thickness variations. If the variation in thickness is excessive, the lace should also be split—the thick sections should be shaved down.

GETTING STARTED

Before the actual braiding can begin, suitable lace must be cut or selected and cut to the length needed for the project. The lace must be pared appropriately for the project and coated with braiding soap. Good preparation is essential for good work.

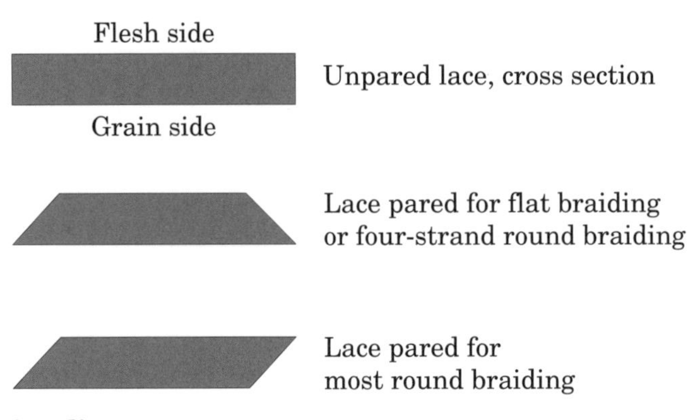

Fig. 6-1. Paring diagram

Length of Lace

The length of lace required for a project may be estimated as one and one-half times the length of the finished project plus a sufficient measure for extras such as fringes, as well as some extra to allow the last part of the braid to be finished with a proper grip on the lace. A bolo tie, for example, with a finished length of 38 inches between the knots, requires 57 inches of lace for the braid, an additional 3 inches at each end for making the tassels, plus about 6 inches for holding the lace as the braiding nears completion, for a total of 69 inches. A hatband with a length of 30 inches plus tassel will require 45 inches of lace for the braid, (doubled to 90 inches if braiding is started with middled strands), plus 3 inches for the fringe (doubled to 6 inches), and another 6 inches may be needed for holding at the end, for a total of 102 inches. In practice, the lengths estimated in this fashion may be generous because of stretch in the lace. In round braids with narrow lace over a heavy core, the estimates may be low. In making a project for the first time it is better to have the strands generous, possibly having to cut off ends that cannot be readily used, than to have strands too short to complete the project. Keeping a record of the lengths of lace used will ensure that the best length for a project is cut when that project or a similar project is next undertaken.

Paring

Lace is pared by drawing the lace past a knife held in the hand. First, the lace should be tied to a hook with a couple of half hitches and stretched (fig. 6-2). Next, a length of leather (about ¾ inch wide and 8 to 10 inches long) should be wrapped around the forefinger of the left hand. The purpose of this leather is to both protect the left forefinger and to give a cradling surface for the lace to slide over. The lace should be run over the protective leather and held against the leather with the left thumb (fig. 6-3a). Left-handed braiders will want to reverse these instructions. With the right hand, the knife is held so as to cut the top left-hand corner of the lace just where the lace leaves the leather wrapped around the forefinger. For stability in holding the knife, the forefinger of the right hand is butted firmly against the forefinger of the left hand. The hands and knife are then drawn steadily away from the hook, allowing the lace to slide through and against the knife-edge. Tension is kept on the lace by the thumb of the left hand.

In paring opposite corners of the lace to make the parallelogram shape used for most round work, the lace is pared on one corner, then

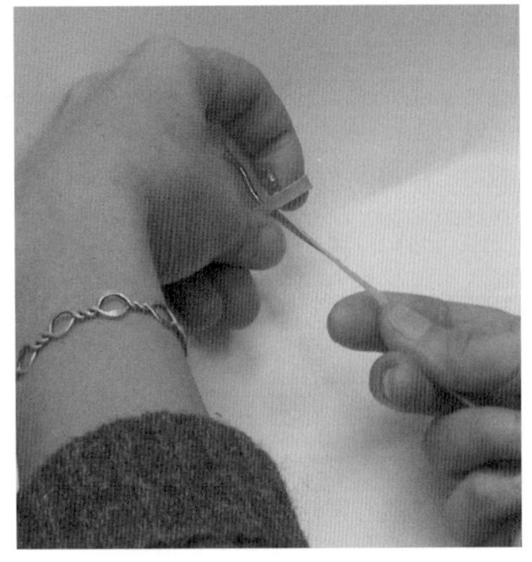

a. Step one: Hold the end of the strand between the third finger and the thumb, form a half hitch by putting a loop over the finger, and put this half hitch on the hook.

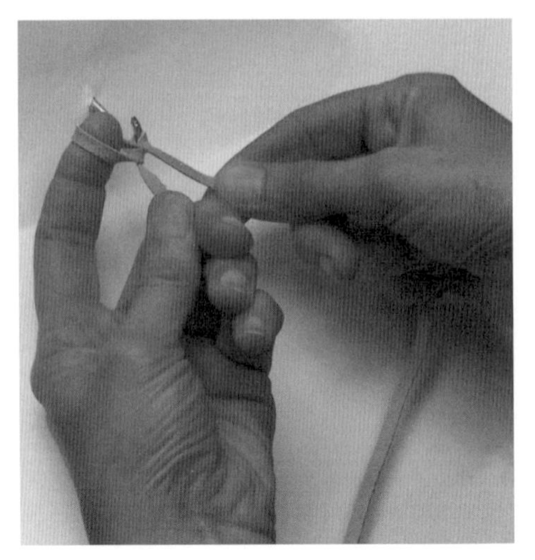

b. Step two: Put another loop over the finger.

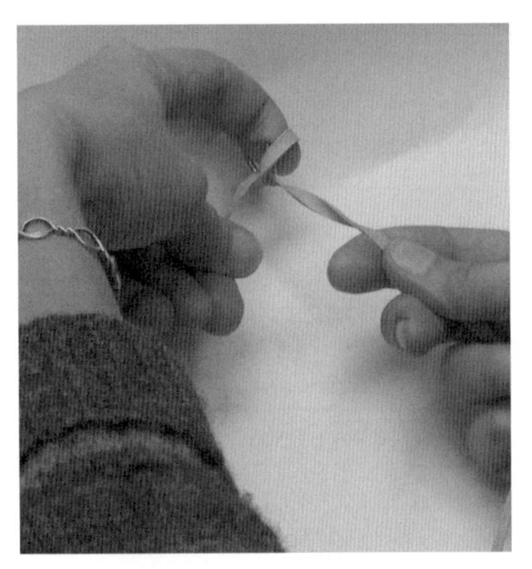

c. Step three: Put this loop over the hook.

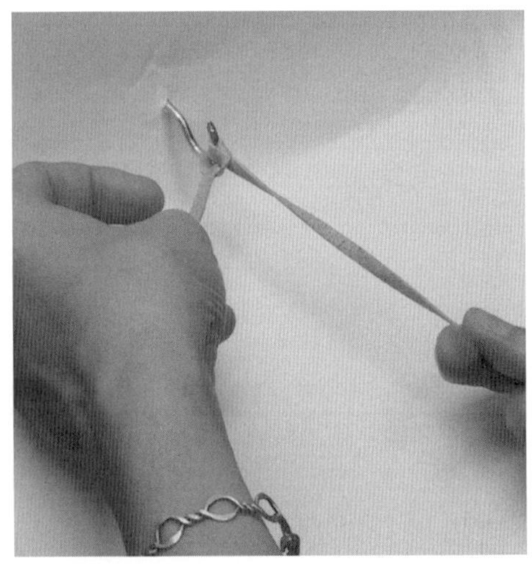

d. Step four: Pull the strand to tighten the knot.

Fig. 6-2. Tying lace on the hook in preparation for paring

turned upside down by twisting near the hook so the opposite corner can be pared.

If it is desired to pare both edges on one side, as for flat or four-strand round work, the knife blade may be held against the left forefinger at a suitable angle to pare the right-hand top edge, again cutting the corner just where the lace leaves the leather over the forefinger. Care must be taken to butt the forefingers of both hands together for stability (fig. 6-3b).

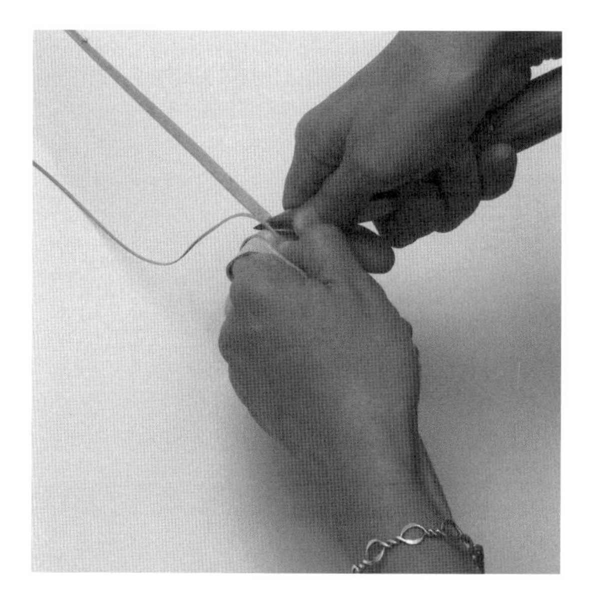

a. Pare the top left corner.

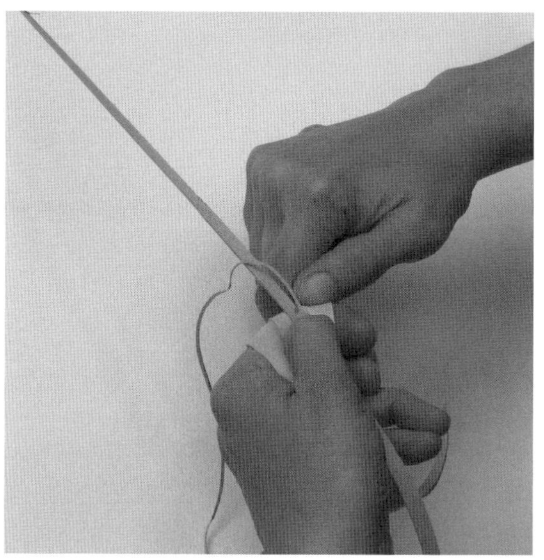

b. Pare the top right corner.

Fig. 6-3. Paring lace before braiding

Braiding Soap

The use of braiding soap is essential for good, tight work. It helps the lace slip into place, and the water in the soap allows the lace to stretch more. This extra stretch, beyond that taken out when stretching the lace before paring, helps the lace adjust to the shape required to fit into a tight braid. The soap should be spread uniformly over both sides of the lace. Cores should be soaped heavily, to ensure that the lace will slip over them easily and also to soften the core to make it more easily compressed (fig. 6-4).

Hatbands should be soaped very little or not at all, as grease on the hat would be objectionable. Talcum powder is usually preferred for braiding hatbands.

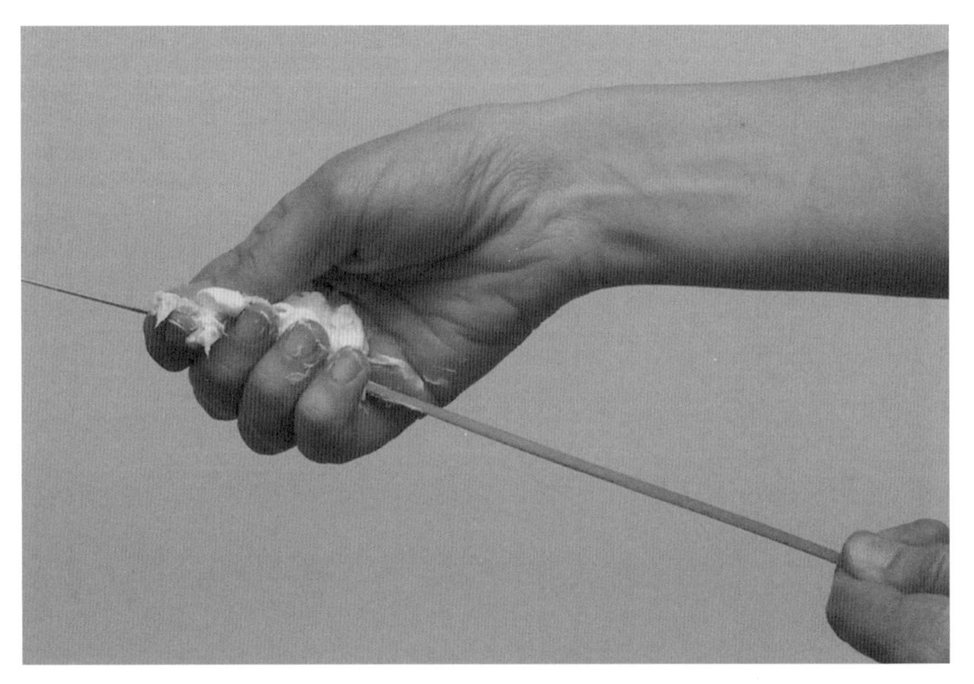

Fig. 6-4. Preparing the lace with braiding soap

CHAPTER SEVEN

Round Braiding

BRAIDING OVERVIEW

Braiding is the placement of strands in their positions in the article being formed. The process may be considered in two sections—first, the pattern of braiding, or how the strands interlock, and second, the mechanics of how the strands are actually put into place. The patterns available range from simple and repetitive to complex and ornate. The choice of pattern affects the appearance of the work, the physical properties of the work, and the ease (or difficulty) in making the braid. By no means do all patterns need to be mastered by the novice braider. However, an early appreciation for the range available is essential for planning projects and for understanding and appreciating the work of others. The mechanics of braiding in a particular pattern should be practiced before attempting to do a project which relies on that pattern. This chapter covers round-braid patterns and chapter 8 covers flat-braid patterns.

THE PATTERNS OF ROUND BRAIDING

A round braid is a form of rope. Just as twisted ropes can vary in the number of strands and the type of twist, so braided ropes can vary in the number of strands and the pattern of braiding. Many variations in appearance are possible because of interplay between patterns of braid and colors of strands. See figure 7-1.

Standard Patterns

Round braids consist of four or more strands divided into two groups which spiral in opposite directions. The spiraling strands interlock as they pass over and under each other. An even number of strands must be used for balanced work, and a number divisible by four must be used for fully symmetrical work.

Four-strand braids are the most simple type. Each strand goes alternately under and over the strands of the opposing spiral. No alternative pattern is practical. The four-strand braid may be formed with or without a core. A four-strand braid with no core may be made firm and round with no hole left in the center. The diameter of the braid with no core will be a function of both the width and the thickness of the lace. The diameter of the braid with a core will be determined mainly by the diameter of the core.

Six-strand braids may be made with each strand going alternately under and over the strands of the opposing spiral. This type of pattern is known as single-diamond work. Alternatively, the strands in one spiral may go under two strands and over one of the opposing spiral, the strands of the latter spiral going under one strand and over two of the first spiral. This type of pattern is known as four-seam work, since four seams run the length of the braid. The single-diamond pattern requires a core for firm solid work; four-seam work in six strands may be made with no core.

Eight-strand round braids are made in either an under-one and over-one pattern (single-diamond), or in an under-two and over-two pattern (four-seam). The first pattern requires a core for firm solid work. The second can be made with no core, but naturally forms a square shape. It may be rolled somewhat round, although if a distinctly round shape is required the use of a core is preferred.

Similarly, braids of ten, twelve, or more strands may be made with the strands going under and over one, two, three, or more strands of the opposing spiral. All should be made over a round core. The sixteen-strand braid may be used as an example to demonstrate the various patterns possible.

The basic pattern for a sixteen-strand round braid has each strand alternately going under four and over four strands of the opposing spiral. This pattern is the one most readily executed, and it forms a smooth round shape with minimum distortion of the leather. It is the ordinary sixteen-strand braid and is referred to as four-seam work in sixteen strands. The seams occur where strands from underneath come to the surface in rows (in this case four) running the length of the braid.

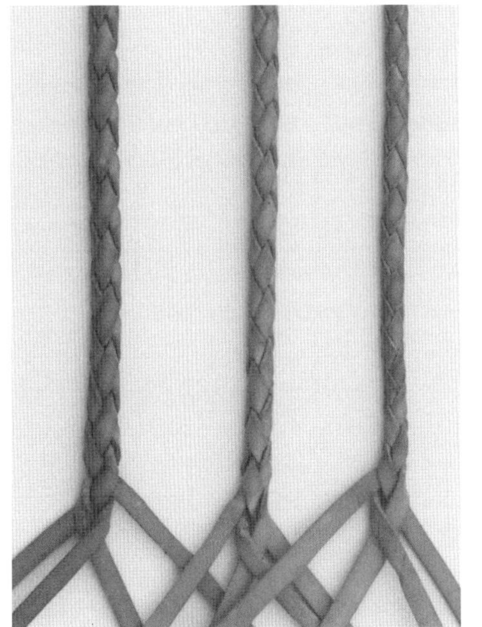

a. Four-strand round braid, no core. Three samples in ⅛-inch lace with different thicknesses of lace are shown.

b. Six-strand round braid, four-seam, no core

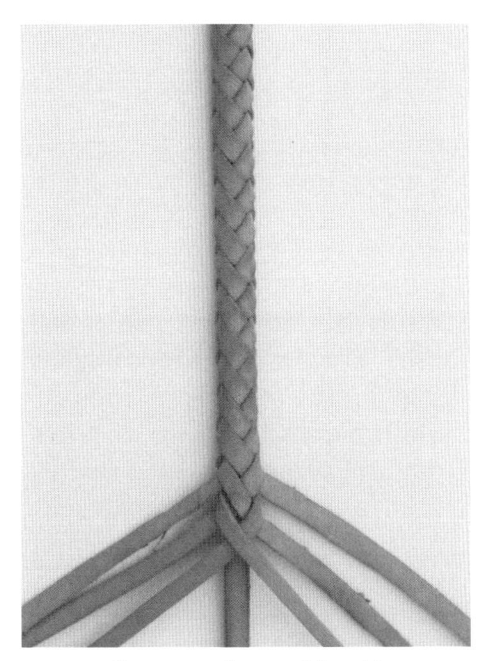

c. Six-strand round braid, four-seam, with core

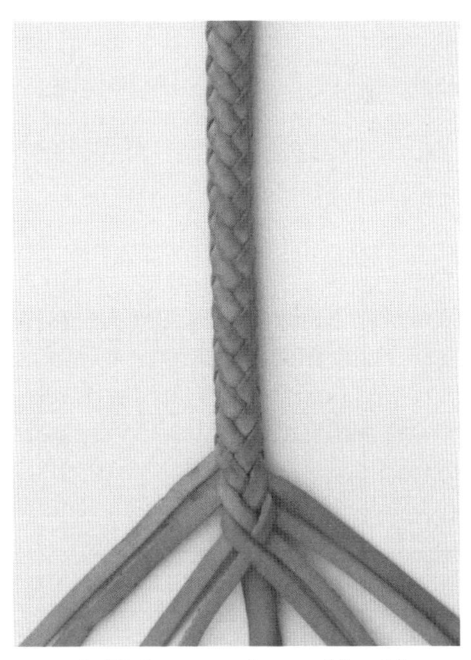

d. Eight-strand round braid, four-seam, with core

Fig. 7-1. Standard patterns of round braids

If the strands are placed alternately going under two and over two strands of the opposing spiral, eight seams are formed rather than four. The braid is called eight-seam work in sixteen strands.

The strands may also go under one and over one of the opposing spirals. Although this forms sixteen seams it is not known as sixteen-seam work, but rather as single-diamond work, or, in American practice, "Dutch plait."

Three patterns—four-seam, eight-seam, and single-diamond—are fully symmetrical and each strand is worked the same way as the previous strand (fig. 7-2). A further series of symmetrical patterns may be formed if multiple strands are worked together. Two strands may be worked as a pair, lying side by side, each pair going alternately under one pair and over one pair of the opposing spiral. This pattern is known as double-diamond work. It is equivalent to single-diamond work done in eight strands, but it is done in eight groups of two strands each.

Similarly, four strands, side by side, may be worked as a group, going alternately under and over the groups of four strands of the opposing spiral. This is known as quadruple-diamond work and is the equivalent of simple four-strand work but is done in four groups of four strands each.

Ring Work

Another series of patterns, generally classed as ring work, may be formed if each strand traces the same path as adjacent strands but in a changing sequence. The seams will run around the braid rather than remaining lengthwise, to show simple rings or the pattern of chevrons known as the "gaucho braid."

Asymmetric Patterns

If no restrictions are placed on how strands go under and over, other than their interlocking in two opposing sets of spirals, a wide range of patterns or designs may be made. By allowing the strands to trace different paths from their neighbors, combinations of regular patterns, repetitions of irregular patterns, or completely freestyle patterns are available. In general, freestyle patterns are restricted to work in two colors. Regular pattern combinations or repetitions of small designs may be carried out successfully in either two-color or one-color work.

Freestyle patterns done in two colors of strands are common in Australia, usually in conjunction with the more symmetrical ring work. Australian ring work normally uses twenty or more strands, although

a. Sixteen-strand four-seam braid

b. Sixteen-strand four-seam braid with two contrasting strands

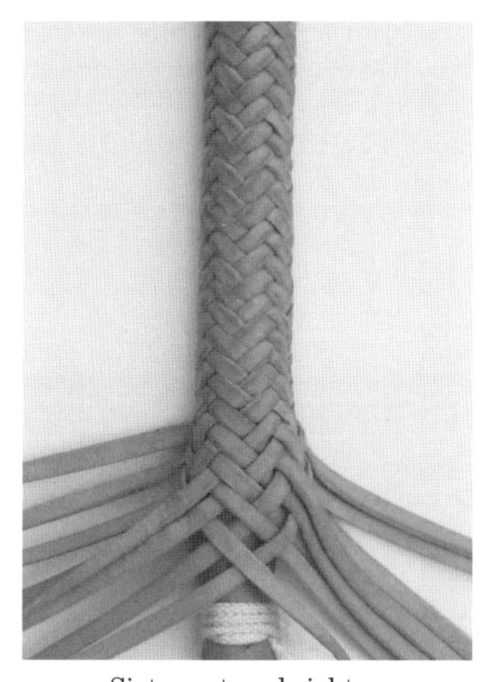

c. Sixteen-strand eight-seam braid

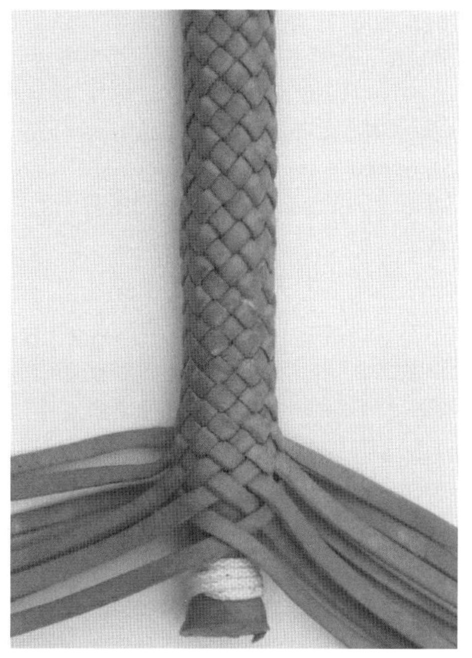

d. Sixteen-strand single-diamond braid

Fig. 7-2. Patterns of sixteen-strand braid

simple patterns are very effective in sixteen strands. All the strands comprising one spiral are one color, while all the strands in the other spiral are a second color. The freestyle pattern is formed by using spirals of the first color as the background. The strands from the opposite spiral (the other color) then show the design. A freestyle design is often set between two rings providing a frame for the design. The central design may be initials, a name, or any other design that may be formed by squares (or diamonds) of contrasting color. Many types of fancy braids are shown in the *Encyclopedia of Rawhide and Leather Braiding* by Bruce Grant and *How to Make Whips* by Ron Edwards.

MECHANICS OF ROUND BRAIDING

The braiding method used by Australian whipmakers allows them to produce firm even work efficiently. This method is presented here in detail.

Attaching the Strands to the Hook

Several methods are used to attach strands to the hook. Middled strands may be put on the hook and the braid started directly, or a loop may be made at the middle of the strands with an overhand knot and the loop put on the hook. The strands may be cut to a yoke and the yoke tied or attached to the hook. Strands not middled can be attached by hitching the ends to the hook. It is difficult to hitch more than four strands to a small hook, but a simple overhand knot may be put on the end of the bundle of strands and the knot placed over the hook with half the strands on each side of the hook.

Four-Strand Round Braid Without a Core

Four ⅛-inch-wide laces about 3 feet long (or two laces about 6 feet long middled with a loop put on the hook) are suitable for practice. The flesh (suede) side of the lace will be on the inside of the braid, and the grain (smooth or shiny) side will be on the outside. First, all strands should be stretched and pared on both flesh corners, then greased with braiding soap. The sequence of setting up the braid may be followed in figure 7-3. Once the initial setup is complete the braiding procedure to produce a firm and even braid with minimum effort can be carried out. The manner of holding the strands and the sequence of operations may be followed in figure 7-4. The hands should be relaxed during braiding except when a strand is pulled. The braid will be somewhat loose at the end, but will tighten up two or three turns back when the strand is pulled.

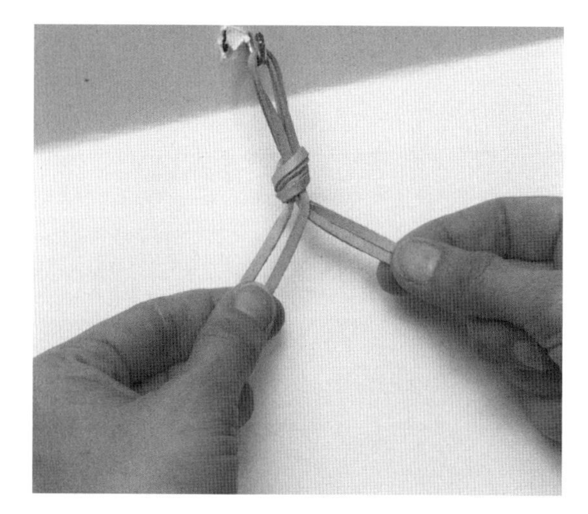

a. Step one: At the knot, cross two of the four strands, right over left, holding two strands slanting out to each side. Hold the two crossed strands grain-side up.

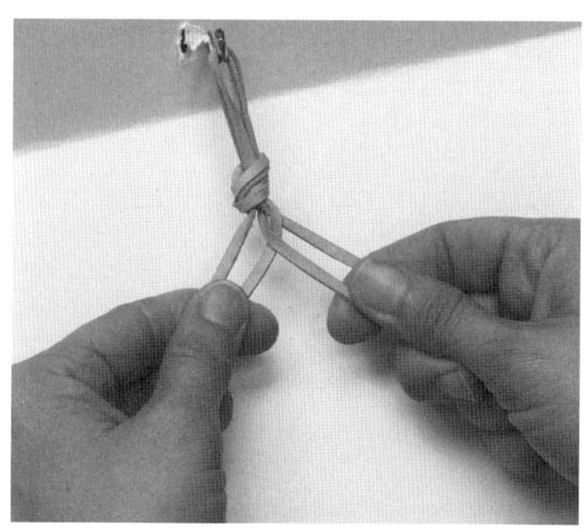

b. Step two: Bring the free (not crossed) strand on the right behind to the left, under the free strand on the left and over the crossed strand. Turn this strand if necessary to make it grain-side up.

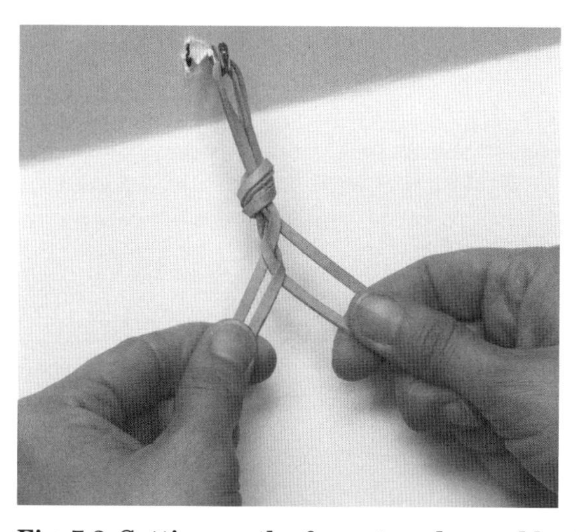

c. Step three: Take the free strand on the left behind to the right, under the upper strand (nearest to the hook) and over the lower strand. Again, make sure this strand is grain-side up. This completes the setup for a four-strand braid. The next stitch will take the strand on the right nearest the hook (the upper strand) around to the left under the upper strand and over the lower strand.

Fig. 7-3. Setting up the four-strand round braid without a core

a. Step one: Start with the center strands crossing right over left. Hold with the right hand where the two center strands are crossed. A tight grip with the thumb is not required. The braid is held with both right strands on the knuckle side of the hand, so that after the working strand has been pulled around and put in place the hand will be in position for the next pull.

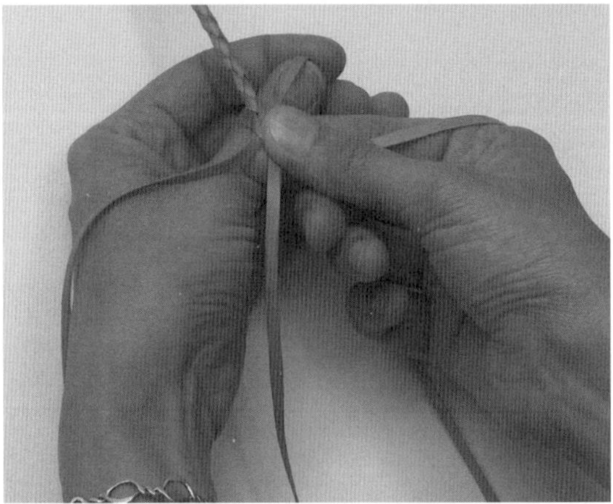

b. Step two: With the left (working) hand, reach between the two strands on the left and pick up the upper right strand (the one closest to the hook). Pick up the strand close to the braid to ensure that it is not twisted.

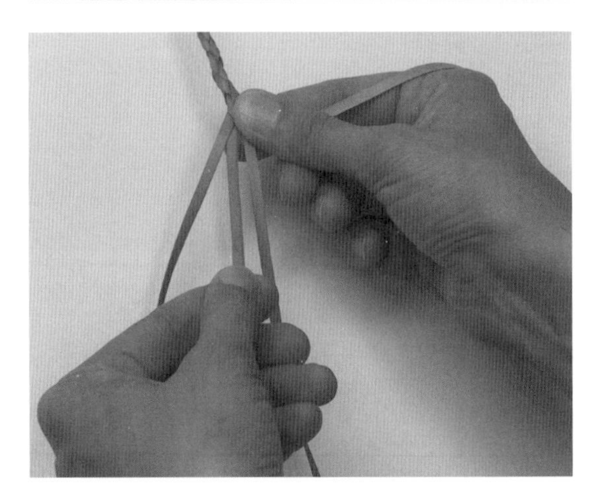

c. Step three: Bring the strand through between the upper and lower strands on the left, allowing it to slip through the fingers to about six inches from the braid. The strand should be flesh-side up.

Fig. 7-4. Braiding procedure for four-strand round braid without a core

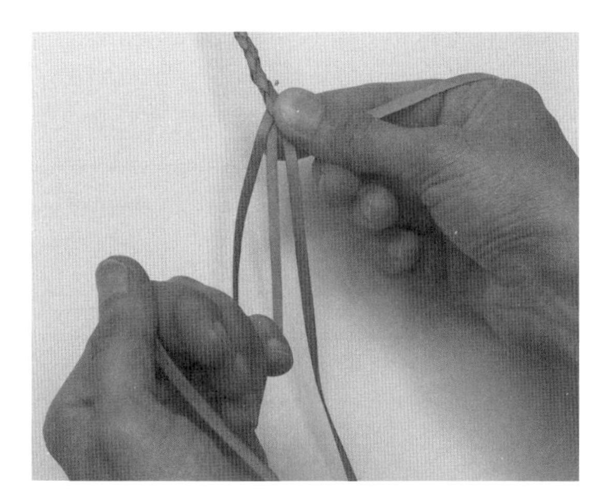

d. Step four: With the left hand, hook three fingers over the strand, thumb holding the strand against the index finger, and pull the strand tight. (The grip is shown in figure 7-5.) The strand should remain flesh-side up during pulling. Relax the hands again after pulling.

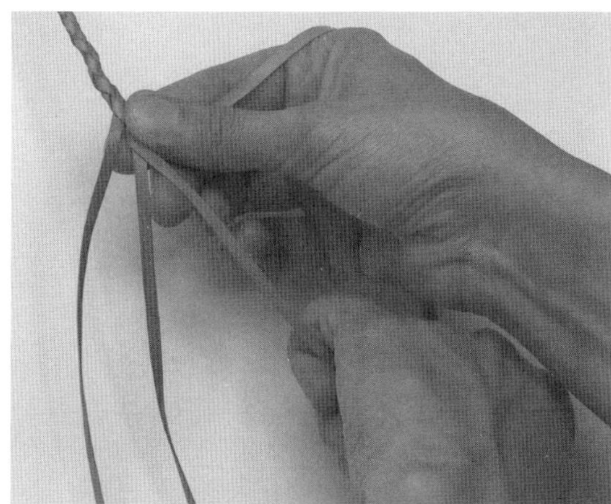

e. Step five: Bring the strand over to the right and lay it in place beside the other right strand. A rolling action with the hand will ensure that the grain side is up when the strand is positioned. The thumb of the right hand acts as a guide for the proper position.

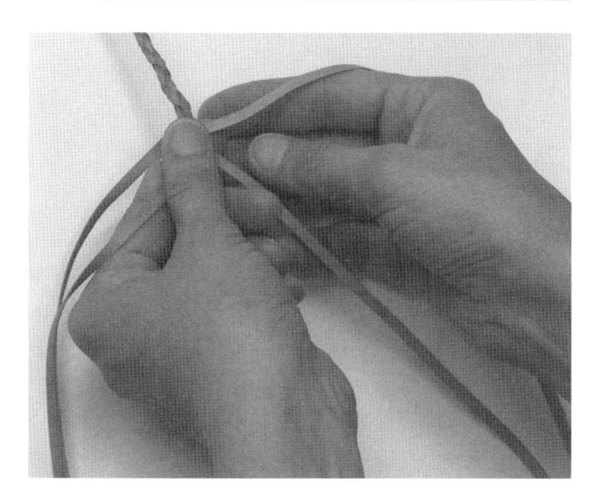

f. Step six: To change to holding with the left hand, slide the right thumb down along the last strand worked and replace it with the left thumb. The left hand will move under the two strands on the left so that they will lie on the knuckle side of the hand. Do not withdraw the right hand, as it is now properly positioned between the two right strands.

Fig. 7-4.—*Continued*

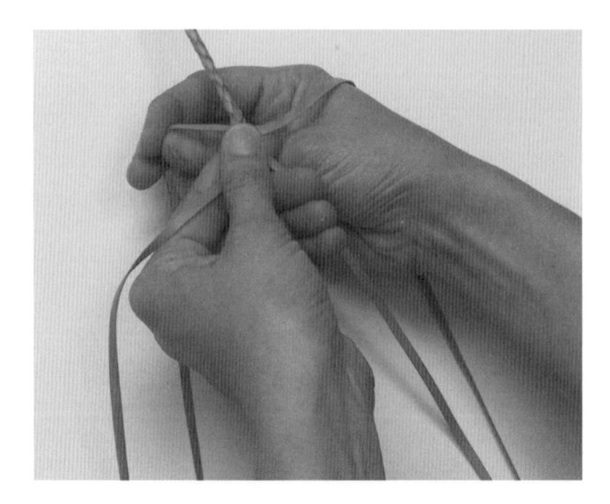

g. Step seven: With the right hand, reach through to grasp the upper left strand between the thumb and index finger.

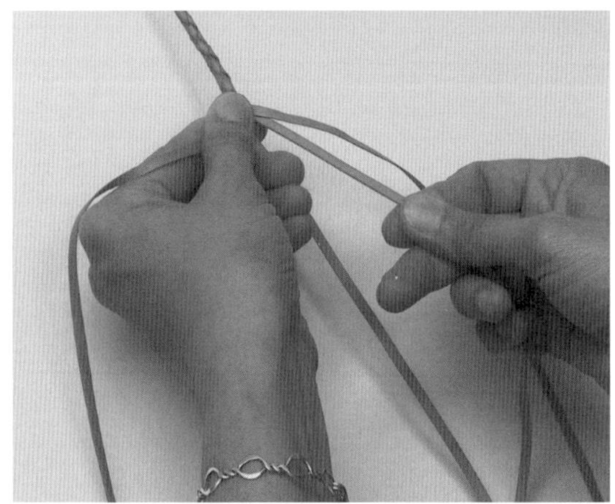

h. Step eight: Bring the upper left strand through between the two right strands and allow it to slip through the fingers to give enough lace to grip properly.

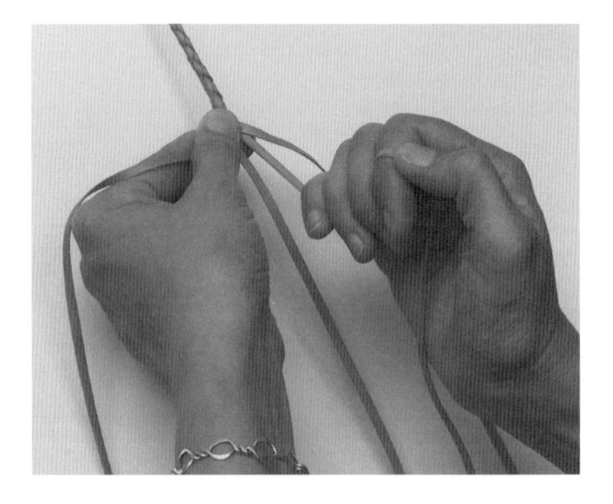

i. Step nine: Grip the strand and pull it tight. Keep the flesh side of the strand up when pulling, and note that when the strand is pulled it tightens the braid farther back. Again, remember to relax the hands after pulling.

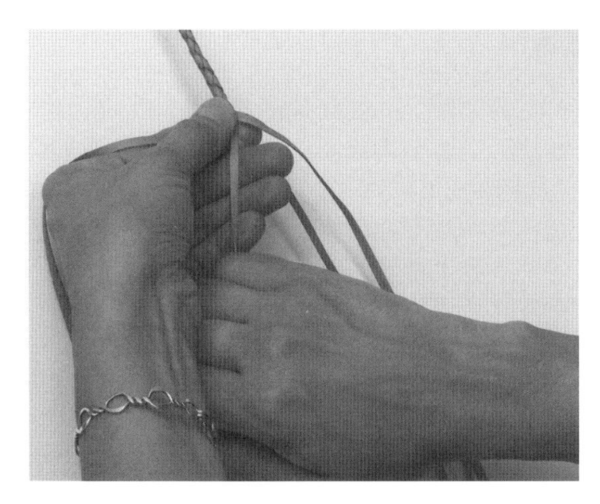

j. Step ten: Lay this strand in place beside the thumb and alongside the other strand on the left. Note that the hand rolls over to ensure that the strand has the grain-side up when it is laid in place.

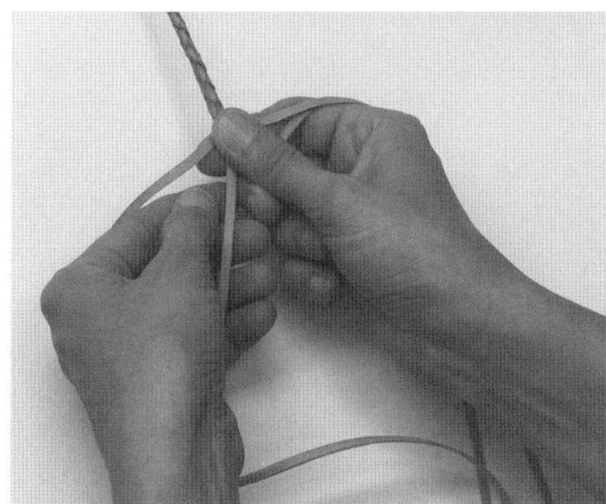

k. Step eleven: Change to holding with the right hand. Note that the left hand is not withdrawn, and that the right hand takes hold with both right strands on the knuckle side of the hand. This completes the full cycle.

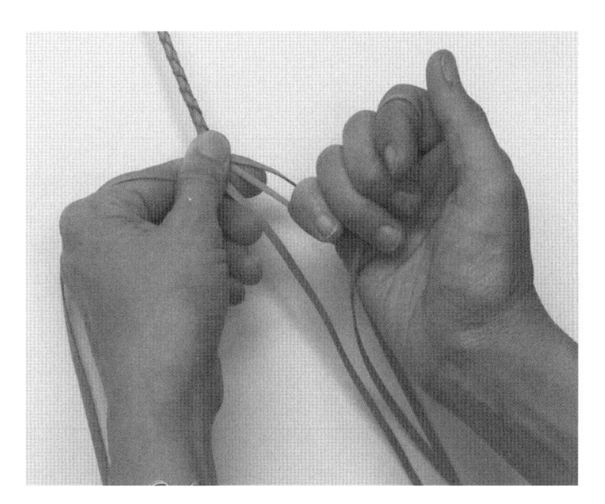

Fig. 7-5. Hand grip, showing the path of the strand when it is gripped by hooking three fingers over it. The thumb will hold the strand tightly against the index finger during pulling.

Stopping and Restarting Braiding

If it is necessary to stop braiding when the work is only partly done, the braid can be kept in order readily by looping the strands from one hand over the hook. If the braid is to be removed from the hook, a slip hitch can be tied on the end to hold the strands together. Use the uppermost strand to tie the hitch (fig. 7-6). If the braid is finished and is to be tied off permanently, all strands should be tightened and a hitch put on the end. See figure 7-7.

It may happen that for some reason the strands become disordered so that the braiding procedure is unclear. In this case the easiest way to sort out the strands is to pick any two that are crossed. Lift these out

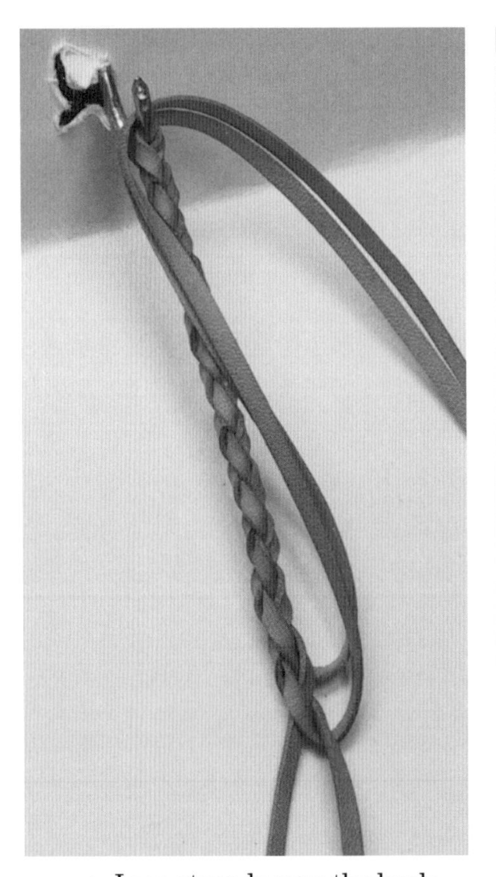

a. Loop strands over the hook.

b. Tie a slip hitch on the end of the braid.

Fig. 7-6. Keeping the strands in order if braiding is stopped temporarily

of the braid, and then take the strands toward the hook and next to them in sequence out to the proper sides (fig. 7-8). The braid will then be in good order.

Fig. 7-7. Hitch on the end of the braid. All strands should be tightened in sequence to tighten the end of the braid before the braid is tied off permanently.

Four-Strand Round Braid Over a Core

Four-strand braids may be made over a stiff or a flexible core. The lace used with a core may be wider for its thickness than lace used in braiding with no core. Optimum widths are determined by experiment. The technique of making the braid is much the same as the one used when there is no core, except for the direction of pull. In braiding with no core the lace is pulled straight from the hook; in braiding with a core the lace is pulled at an angle to the axis of the braid. Pulling at an angle ensures that each strand lies closely beside its neighbor. The strand is pulled against the thumb of the holding hand. The core lies across the palm of the hand holding the braid during working, more or less directly in a line from the hook to the braider.

The core can be started at any stage by holding it in place under the working strands. Figure 7-9 shows the sequence of braiding as well as the position of the hands as they pull the strands into place to keep them close together in the braid.

Eight-Strand Four-Seam Round Braid Over a Core

The general case for round braids can be most readily shown with an eight-strand braid done over a core. The width of the strands must be correlated with the diameter of the core. As an approximate rule, the width of the strands may be taken as four times the core's diameter

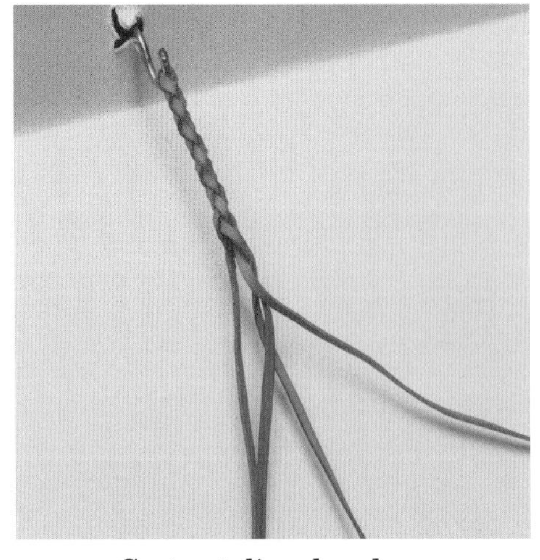

a. Sort out disordered four-strand braid.

b. Select any two crossed strands.

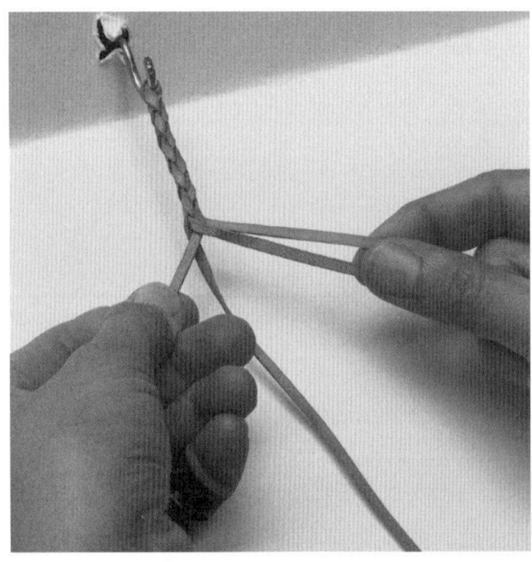

c. Take adjacent strand (nearest to the hook) to the right.

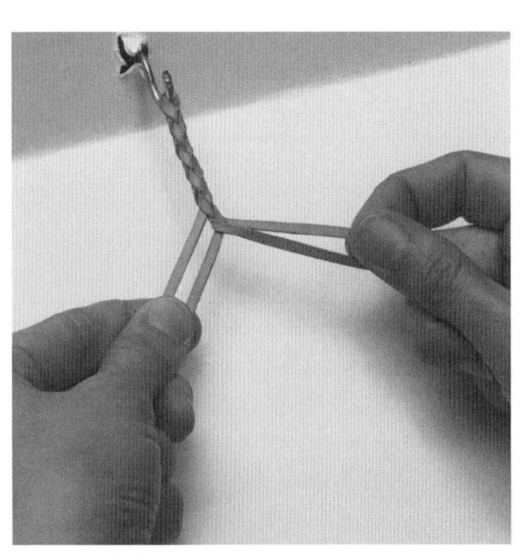

d. Take next strand to the left, and all strands will be in place.

Fig. 7-8. Repositioning strands in proper order

divided by the number of strands. For example, on a ¼-inch diameter core, for eight-strand work the strands should be about ⅛ inch wide (4 multiplied by ¼ and the result divided by 8). The ideal width braids easily and covers the core with no spaces between the strands and without crowding the strands. If the strands are too narrow it is difficult to keep them close together. If the strands are too wide they will appear distorted from crowding. There is a reasonable range of variation in width for an acceptable braid. The preferred width for any specific core and lace can best be determined by experiment.

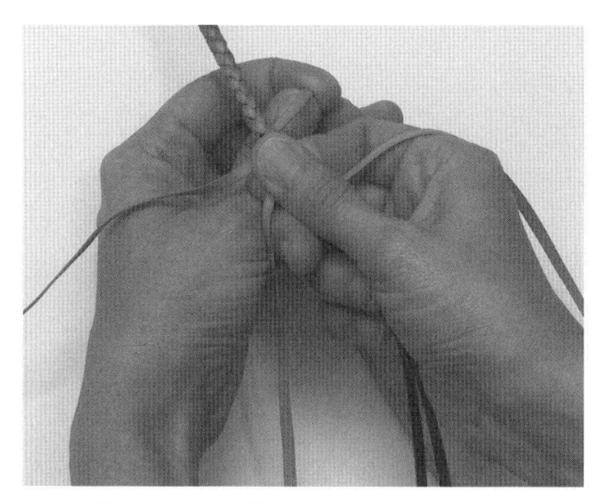

a. Step one: Start with the two center strands crossed right over left. Hold with the right hand, both right strands on the knuckle side; the core, here two black strands, runs across the palm. With the left hand, reach through between the two left strands to pick up the working strand.

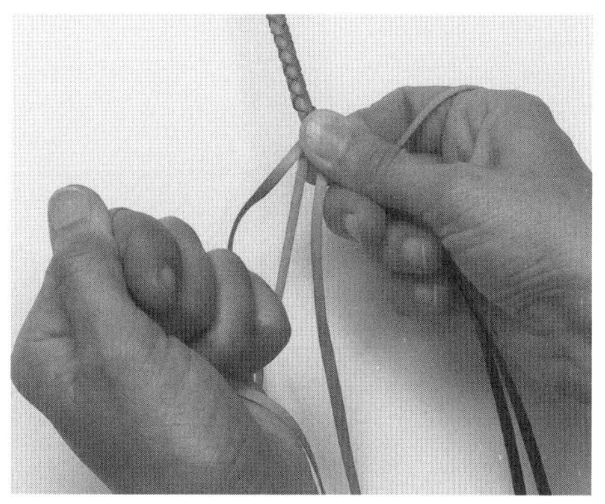

b. Step two: Grip the strand with the left hand. The pull would be at the angle shown only with strands very wide for the core. With narrower strands, the pull would be at an angle to the axis of the braid, as shown in fig. 7-9c.

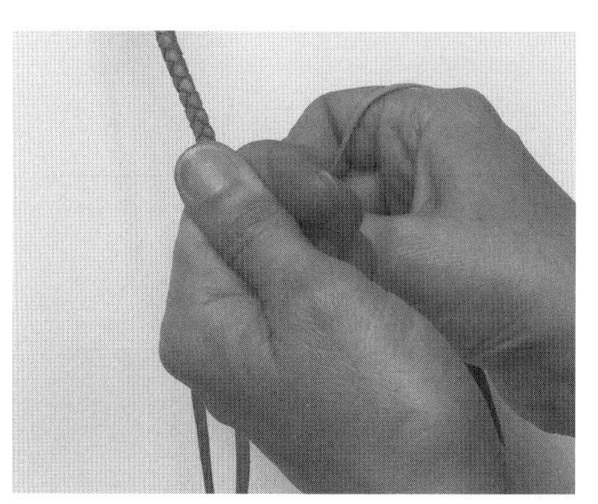

c. Step three: The left hand pulls the working strand. Top view as seen in braiding.

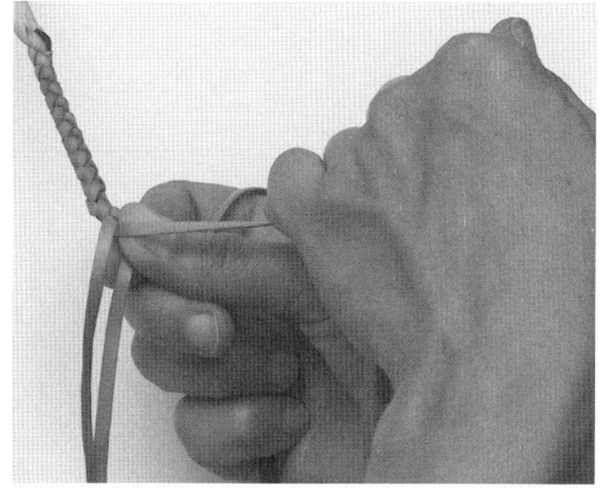

(Side view of step three, showing how the strand is pulled at an angle against the thumb to keep the strands closed up.)

Fig. 7-9. Braiding procedure for four-strand round braid over a core

In an eight-strand braid, four strands spiral to the left and four spiral to the right. For normal four-seam work, each individual strand alternately passes under two and over two strands of the group spiraling in the other direction, thereby forming a chevron or herringbone-like pattern. For a practice braid, eight strands can be placed on the hook. They can be knotted, middled and knotted with a loop, or cut to a yoke.

Fig. 7-9.—*Continued*

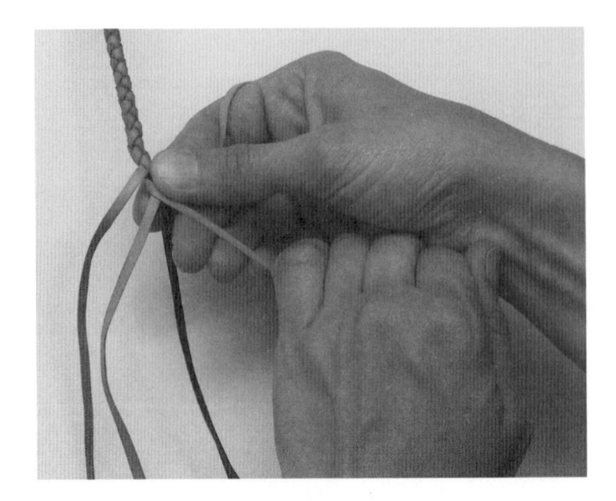

d. Step four: With the left hand, place the working strand adjacent to the other strand on the right, close under the thumb.

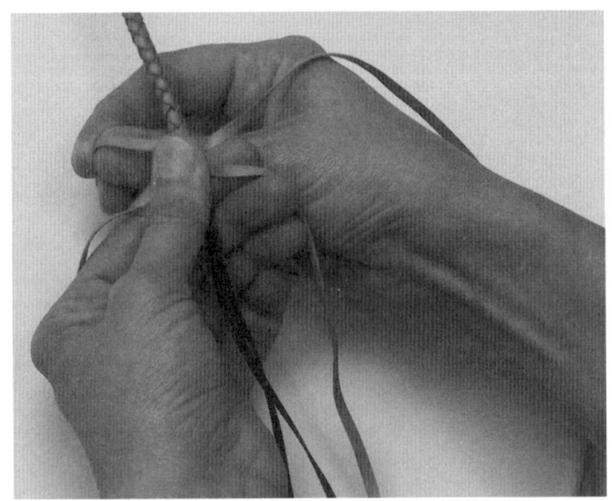

e. Step five: Change to working with the right hand. The left hand takes up the braid with the two left strands on the knuckle side and the core running over the palm.

All braiding strands should be stretched and pared on opposite corners, then greased with braiding soap. The sequence in setting up the braid is shown in figure 7-10. The position of the hands and the sequence of operations to produce a firm and even eight-strand braid are shown in figure 7-11. Two similar strands can be tied to the hook to be used as a core. To help keep the strands close together in the braid, each strand being braided should be pulled tight at an angle against the thumb. The hands should be relaxed except when pulling a strand tight.

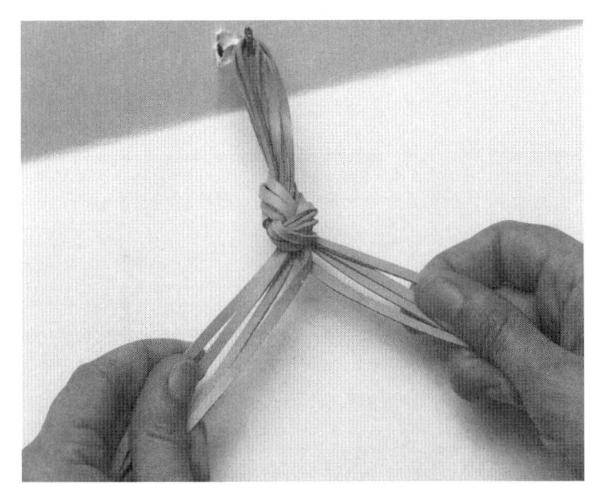

a. Step one: Put eight strands on the hook, middled and knotted. Divide the strands into groups of four to the right and four to the left. Where convenient, have the strands grain-side up. Cross two in the center, right over left, with the grain-side up.

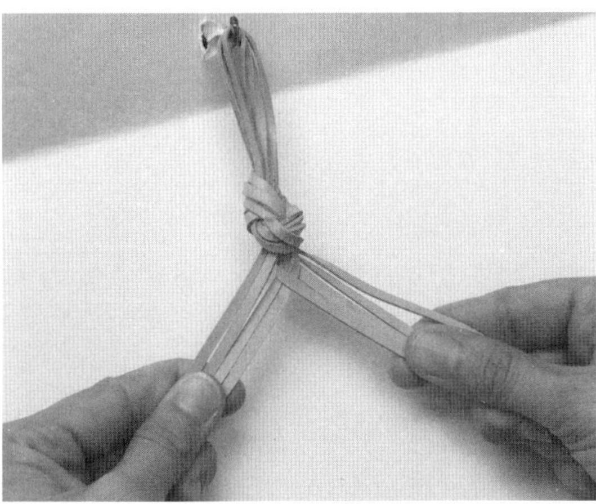

b. Step two: Take the upper strand on the right around the back under the two upper strands on the left and over the two lower strands. Lay it in place with the grain-side up.

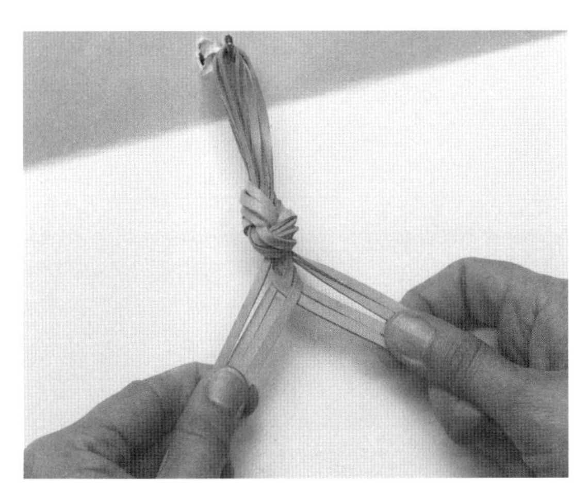

c. Step three: Take the upper strand on the left around the back under the two upper strands on the right and over the two lower strands. Again, keep the grain-side up.

Fig. 7-10. Setting up the eight-strand four-seam round braid

Fig. 7-10.—*Continued*

d. Step four: Take the upper strand on the right around the back under the two upper strands on the left and over the two lower strands.

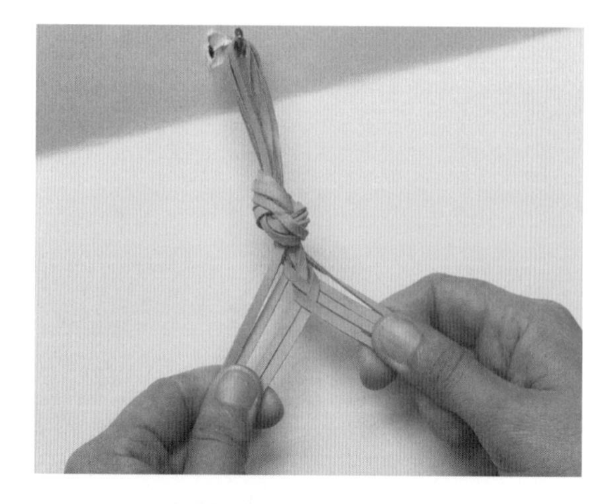

e. Step five: Take the upper strand on the left around the back under the two upper strands on the right and over the two lower strands. These steps set up the pattern of braiding. The next strand will be taken from the right around the back under the two upper strands on the left and over the two lower strands. In setting this up, make sure that all strands are grain-side up.

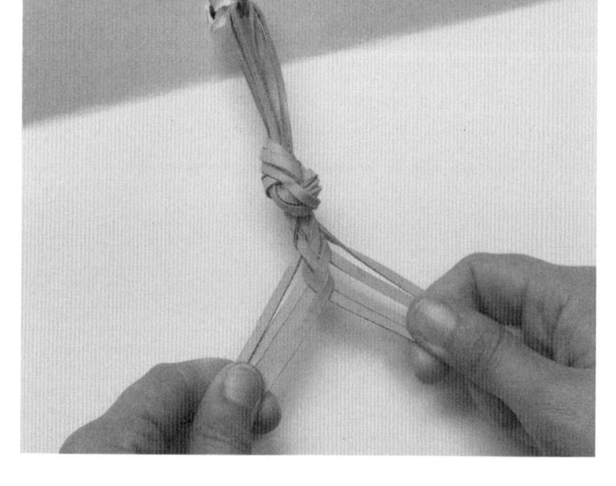

f. Each stitch follows the same path.

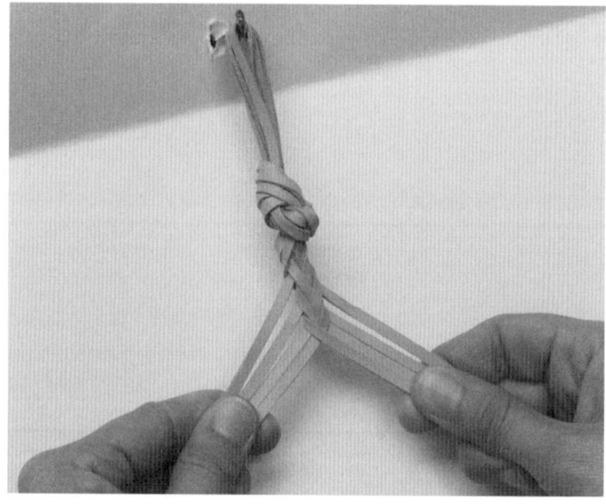

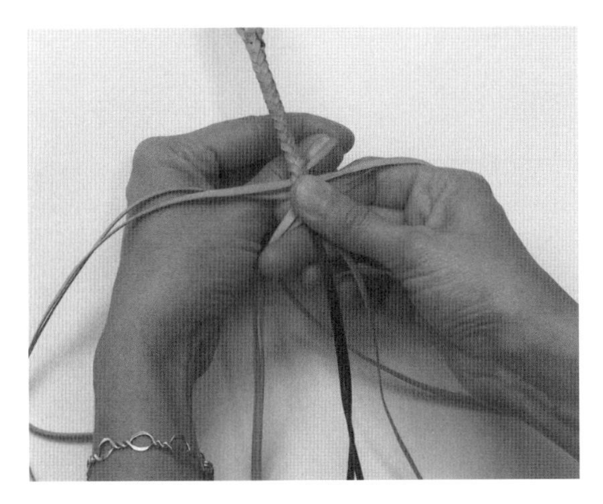

a. Step one: Start with the center strands crossed right over left. Hold with the right hand, the three upper strands on the right on the knuckle side of the hand, the lower strand and the core on the palm side. The thumb should be on the cross. Reach through under the two upper left strands and over the two lower to pick up the upper right strand.

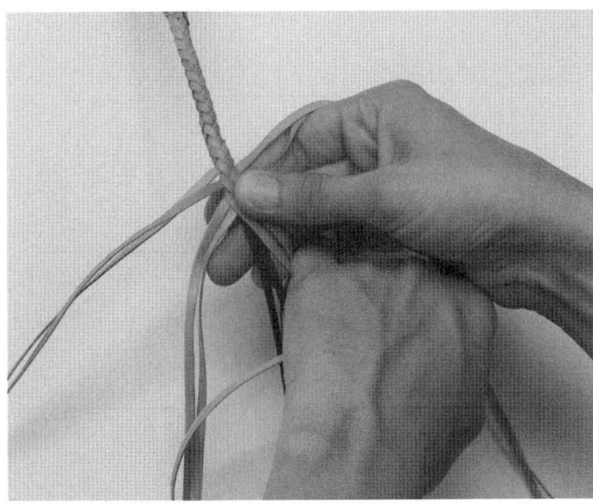

b. Step two: Bring the upper right strand through, pull it tight at an angle against the thumb, and lay it in place. Note that this leaves two strands on the knuckle side of the right hand and two strands on the palm side.

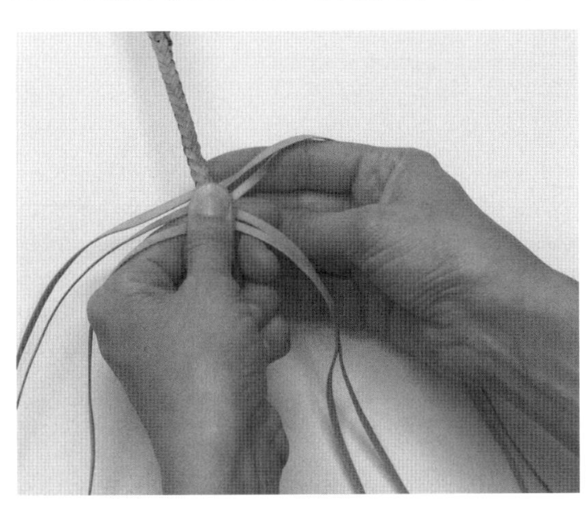

c. Step three: Change to holding with the left hand. The three upper strands on the left will be on the knuckle side of the hand. Do not withdraw the right hand, as it is now positioned between the upper two and lower two strands, ready to pick up the next working strand from the left.

Fig. 7-11. Sequence of operations to produce an eight-strand four-seam round braid over a core

Fig. 7-11.—*Continued*

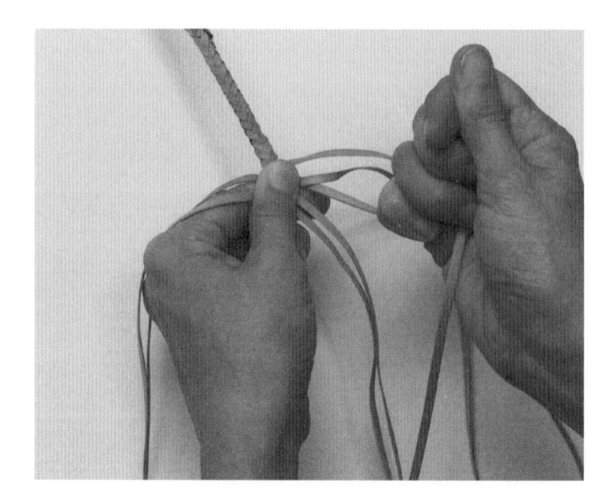

d. Step four: Bring the upper left strand through and pull it tight. The strand is held flesh-side up as it is pulled. Note that the braid stays somewhat loose at the end but tightens up further back when the stitches are pulled.

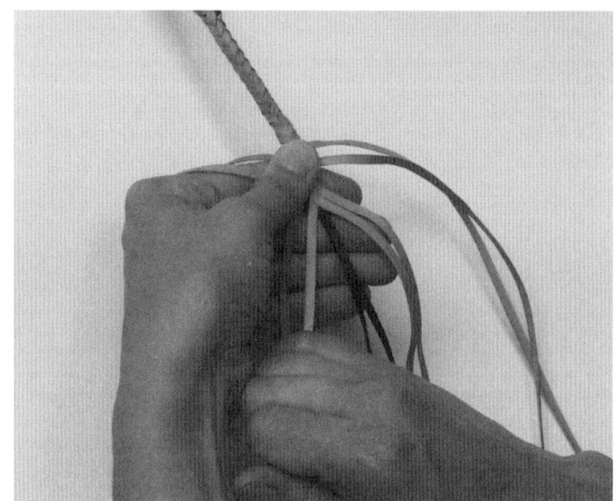

e. Step five: Bring the working strand over and into place with a rolling action to ensure that the grain side is up.

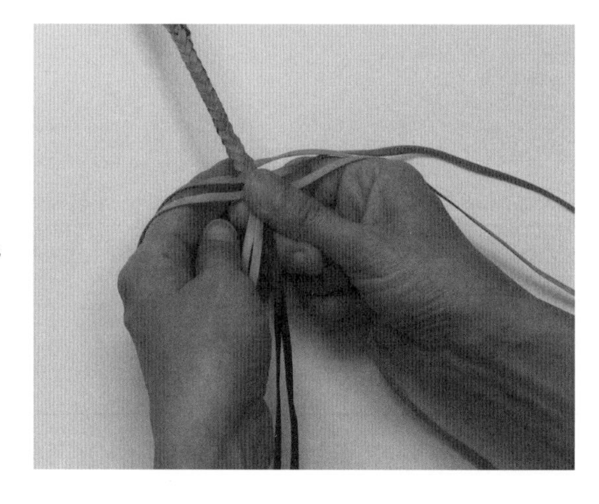

f. Step six: Change hands to hold with the right hand. Note that the three upper strands are on the knuckle side and the lower strand is on the palm side of the hand. The left hand is not withdrawn, since it is in proper position to reach for the next working strand. This completes the full sequence of braiding.

Disordered Braid

If for some reason the strands get mixed up, they can be put in order in the same manner as that used for the four-strand braid. Pick any two strands that are crossed, lift them out, and hold them. In sequence, take the strands adjacent to them on the side toward the hook and bring them out to the proper side. See figure 7-12.

Untangling Long Strands

When strands are braided together, the free ends of the strands are also braided. When the strands are short, the free ends usually swing free

a. An eight-strand braid has become disordered.

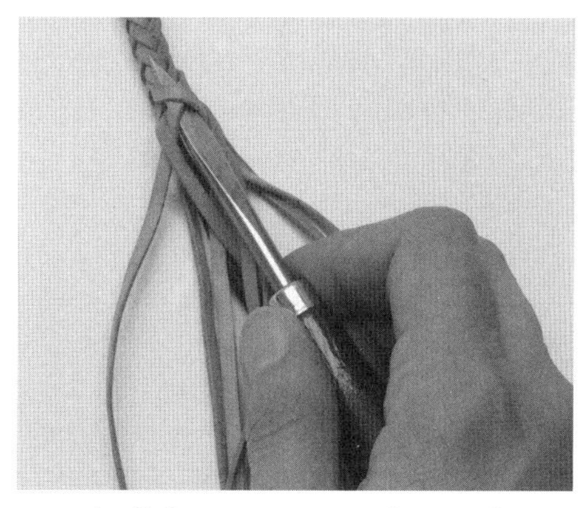

b. Select two crossed strands.

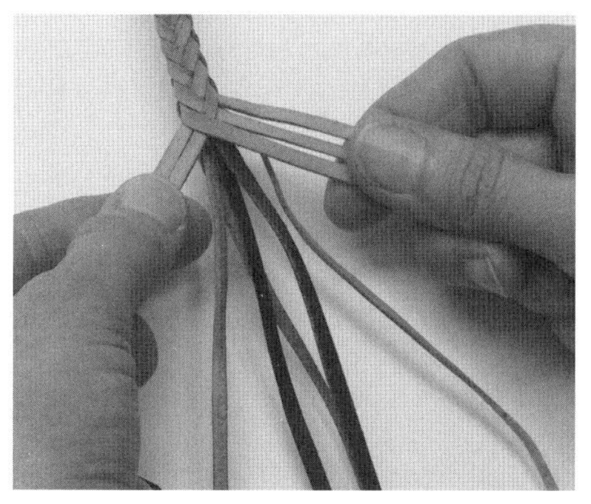

c. Bring the two crossed strands out and hold them in position. Bring the adjacent strands into position in sequence and hold them in the hands.

d. All strands are in correct position.

Fig. 7-12. Repositioning the strands in a disordered eight-strand braid

and work out this secondary braiding, but when the strands are long the free ends will frequently tangle. The tangle should be removed when it interferes with the braiding. To remove the tangle, keep the braid in the left hand as during braiding; hold the top of the tangle with the fingers of the left hand and draw out the strands one at a time with the fingers of the right hand. Figure 7-13 shows the strands being drawn out. All strands should be drawn out including the last, as the twist in the strands is removed at the same time.

Round Braiding in Other (Even) Numbers of Strands

The method of braiding any number of strands in a four-seam round braid may be readily adapted from the method used with eight strands. For a twelve-strand braid, for example, the working strand will be brought under three and over three strands. The hand holding the braid will do so with two strands on the palm side and four on the knuckle side, to be in the proper position (three on each side) when it is time to work the next strand.

For a six-strand braid, where the strands on each side total three, the working strand goes under two and over one on one side, and under one and over two on the other, to produce maximum symmetry. The pattern of the six-strand braid is shown in figure 7-14. In braiding, the right hand holds with all three strands on the knuckle side, the left with two strands on the knuckle side.

Braiding in Single Diamond

Single-diamond braiding, in which each strand spirals in an over-one, under-one pattern, is a particularly useful pattern for the beginning and end of work (fig. 7-15). Because each strand moves no more than the width of a single strand before it is interlocked with another, the single-diamond pattern provides the strongest foundation for back-braiding and a good start to a round braid. It also provides the neatest transition to flat braiding. Thus, the single-diamond pattern is found in the dog-lead project at two places: at the snap, where the braiding starts and the firmness of single-diamond work is needed; and at the end of the round braiding, where the single-diamond pattern provides a better foundation for back-braiding the wrist loop into place and also creates a neater transition to the flat braid of the wrist loop. However, single-diamond braiding is seldom used throughout an entire piece. First, it is more time-consuming. In four-seam work, the holding hand can simply be inserted in the strands on one side in such a way that with a single

a. When a tangle interferes with braiding, it should be removed. Note that the strands in the tangle are braided, and are twisted from the braiding.

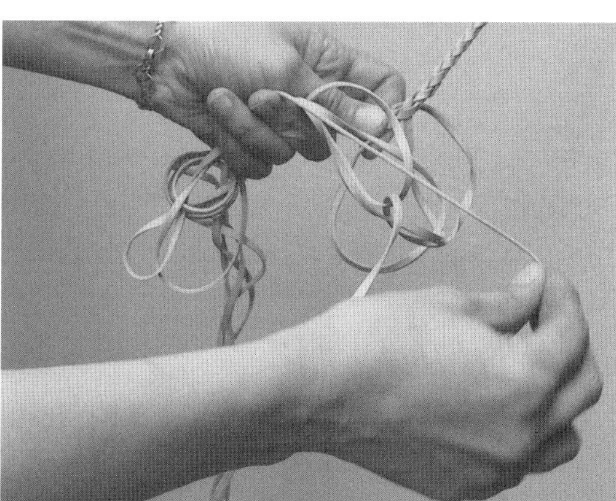

b. Draw out first strand. As it is drawn out from the tangle through the fingers of the left hand, the twist is also removed.

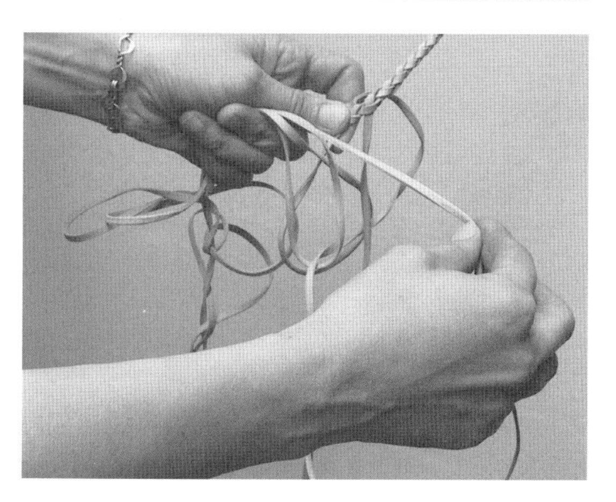

c. Continue to draw out strands, including the last strand, in order to remove the twist in this strand.

Fig. 7-13. Removing a tangle from the free ends of long strands

motion, it is positioned so that it is all set to reach through in the next stitch. Single-diamond work (in anything more than four strands) requires that alternate strands must be sorted out before picking up the working strand, which takes significantly more time. Second, single-diamond braiding is less flexible than four-seam braiding. When four-seam work is bent, the strands on the inside of the curve can partially overlap, and those on the outside can separate. In single-diamond work, the strands cannot overlap as much nor separate as readily.

The techniques used in single-diamond work to sort the strands before bringing through the working strand can be adapted to all the other fancier round braiding patterns, from eight-seam work to ring work and free-form work.

Single-diamond braiding is accomplished with many of the same considerations used in four-seam work. Figure 7-16 shows the procedure for braiding in an eight-strand single-diamond pattern. The steps are readily adapted to six, ten, or more strands. The braid should be started with a standard eight-strand four-seam braid, with the last two strands worked crossed right over left.

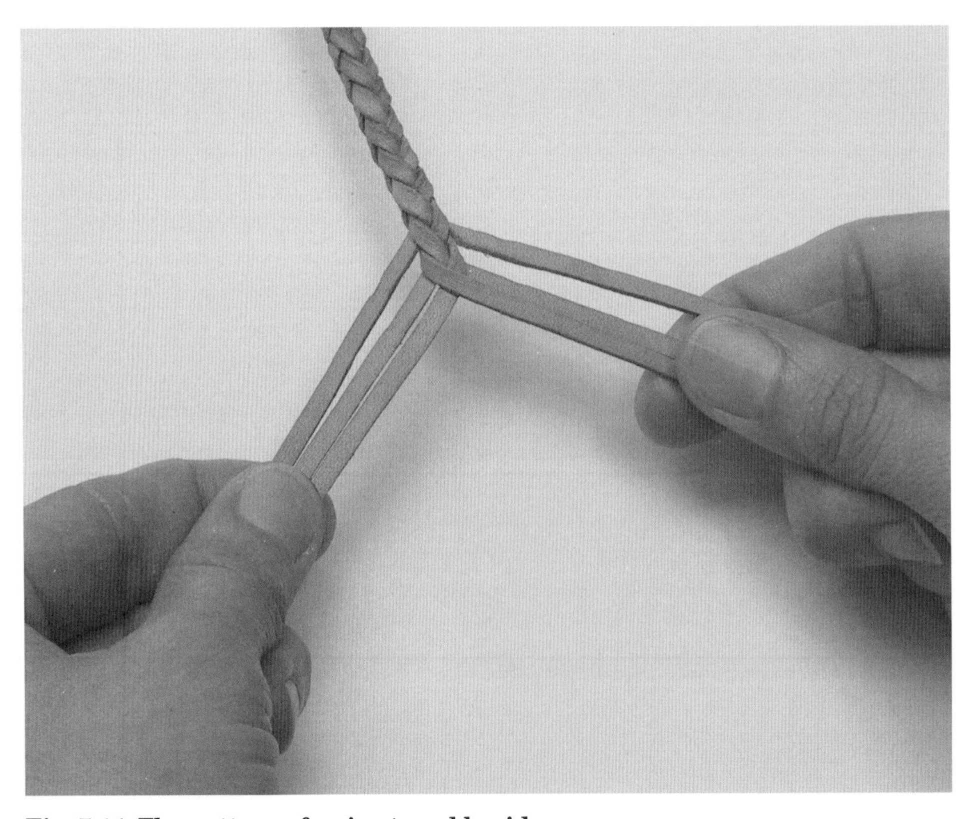

Fig. 7-14. The pattern of a six-strand braid

Fig. 7-15. A single-diamond braid for the handle of a stock whip done in white and brown strands to show up the pattern

Width of Strands

An immediate practical matter to be determined before braiding is the appropriate strand width to be employed. Four- and six-strand round braids may be made without a core. In theory, the width of the strands in a four- or six-strand round braid with no core should be exactly that width at which there is just sufficient leather in the strands to form a solid braid. If the strands are too narrow, there will be gaps between adjacent strands on the surface of the braid. If the strands are too wide, either the braid will be hollow or the strands will be squeezed and crumpled, effectively making them narrower. The appropriate width is then dependent upon the thickness of the leather employed. In practice, because the leathers employed in braiding can be compressed or stretched appreciably, there is a fair range of widths that can be satisfactorily braided in either four- or six-strand round braid without a core. The effect of different widths of lace in these braids can be demonstrated readily with a few samples of braid made with a range of strand widths. As a general start, ⅛-inch lace cut from a medium-weight kangaroo skin will form a good firm four-strand round braid.

Braids made over a core allow a considerably wider range for the width of lace, as the strands can form faster or slower spirals, that is, there will be more or fewer turns per unit length of core. In practice, overly wide lace will produce a braid that appears to be crowded and unattractive, and extremely narrow lace presents problems in braiding.

a. Step one: Hold the crossed center strands with the right hand, one strand on the knuckle side and three strands on the palm side.

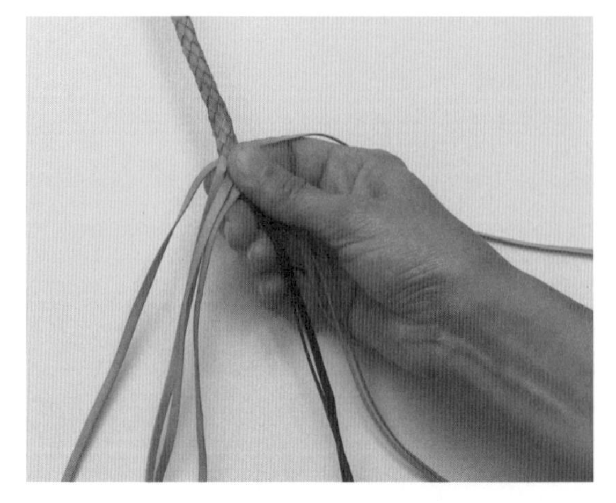

b. Step two: With the left hand, lift the first (upper) and third strands on the left side and depress the second and fourth (lower) strands, holding these with the index finger of the right hand snug against the core. The working strand will pass over these two strands and under the two that are raised.

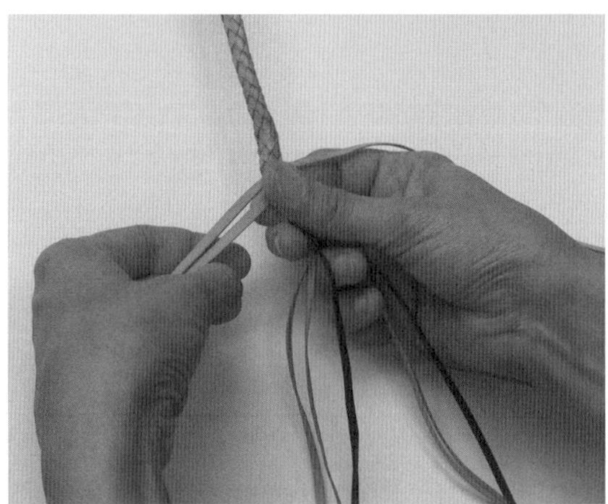

c. Step three: With the left hand, reach through underneath the two raised strands and over the two depressed strands around the core to the uppermost right strand and pick it up close to the braid. Pass it around the core, under the two raised strands, and over the two depressed strands in an under-one, over-one, under-one, over-one sequence.

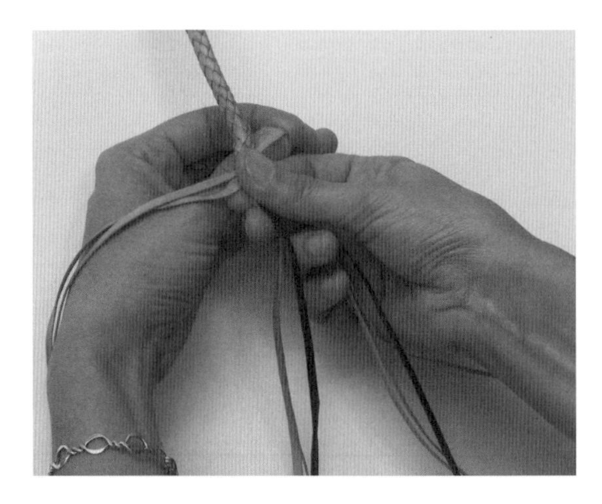

Fig. 7-16. Procedure for braiding an eight-strand single-diamond pattern

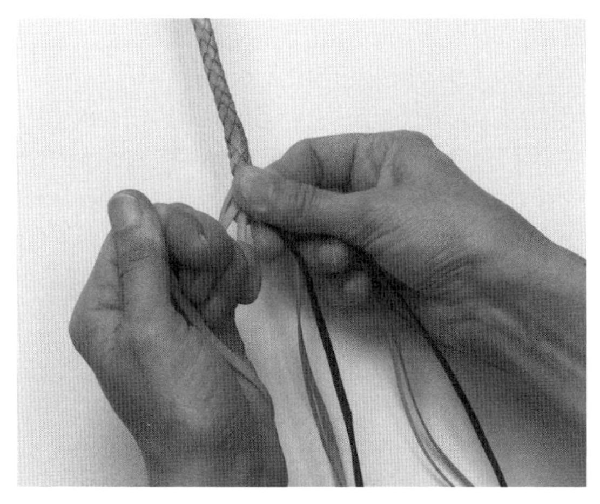

d. Step four: Bring the working strand through and pull it tight, holding it at a slant against the thumb to keep the strands close together in the braid. Hold the strand flesh-side up when pulling.

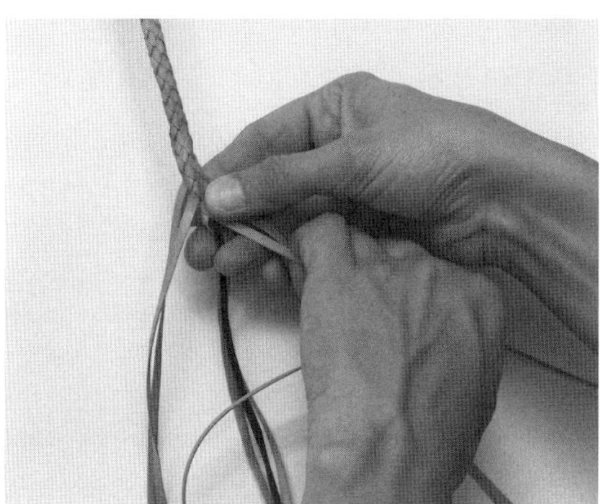

e. Step five: Lay the working strand in place, using a rolling action to keep the grain-side up. This completes the first half of the cycle, in which the braid is held in the right hand and the strands are sorted and worked with the left hand.

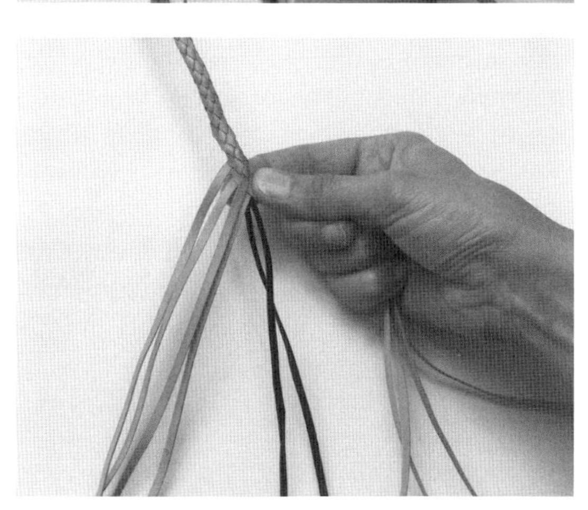

f. Step six: The under-over pattern of the stitch is clearly seen in this illustration. To complete the braiding cycle, hold in the left hand, one strand on the knuckle side, repeat the sorting procedure with the right hand, and bring the upper left strand around and into place.

Strands cut to a width such that the angle of the spirals is 45 degrees to the axis of the braid—or slightly narrower—offer a good compromise to the various factors of appearance, rate of braiding, flexibility, and durability. As a rule of thumb, one may estimate the width of a strand by taking four times the core diameter and dividing this by the number of strands in the braid. Some allowance must be made for loss of width during stretching and paring, and also for the thickness of the leather if the core is relatively small. Here again, some experimentation with various widths of lace on a constant core diameter will readily demonstrate the points indicated above. Note that tension is important, and test the effect of greater or less pull on braids with wider and narrower strands.

LEATHER CORES FOR ROUND BRAIDS

The leather core for round braids must be uniform so that the braid is of uniform diameter, and it must be cut so as to take up a round shape readily. It should be made of a soft leather in order to ensure that a dense core is formed during braiding.

Where the core is relatively small the shape of the leather in the core is of little importance, as the pull of the strands will force it into a satisfactory shape and density. For example, the core for a six-strand braid can be made using a strand of the same size as the braiding strands, with a fully satisfactory result.

Cores may also be made of many very narrow strips. Cores should not, however, be made of a pile of flat strips cut so that they form the core as a rectangle of flat layers. Such a core will compress unevenly, and it will bend more easily in the direction across the thickness of the layer sections than it will in the direction of the width of the layer sections.

The core should be compressed into a round shape, with the layers concentric. Rounding the core before braiding does not set it permanently into this compressed shape, but helps establish the shape so that the core will be compressed into this shape during braiding.

Cores are rounded by dressing them with braiding soap and running a looped strand of lace back and forth over them. For a core made of two strips to form concentric circles, round the inner layer first, then the outer layer with the inner layer in place. Before rounding, the core should be well coated with braiding soap (fig. 7-17).

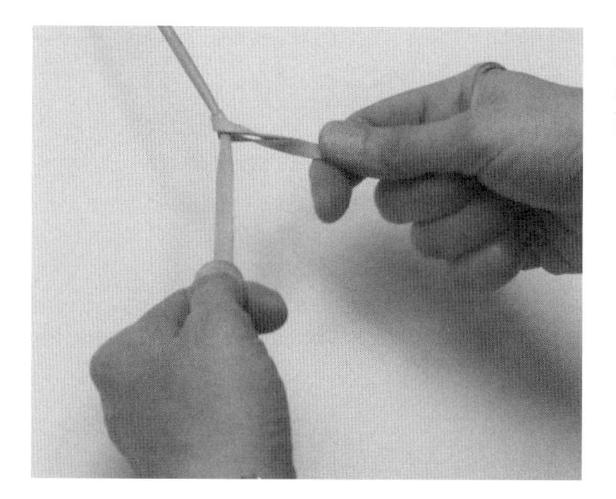

a. Round the inner layer. The core should be well coated with braiding soap.

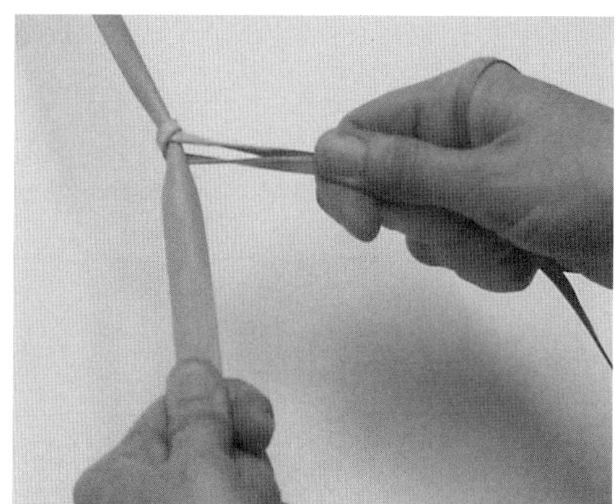

b. Round the outer layer around the already rounded inner layer.

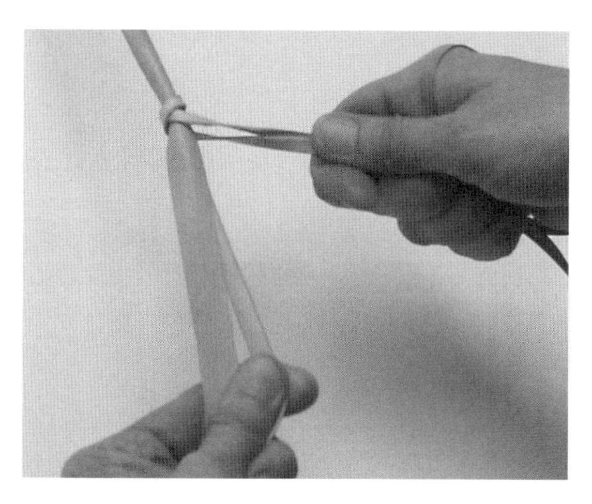

c. The two layers are shown separated beyond the rounded part.

Fig. 7-17. Rounding a core with two layers

a. Eight-strand braid over a two-layer core

b. Cross section of braid

c. Same cross section with braided overlay slightly separated from core

Fig. 7-18. Round core with two layers

Larger cores for round braids may be built up with two or more layers to increase the diameter. An eight-strand braid over a two-layer core is shown in figure 7-18.

Rolling Round Braids

Round braids are usually rolled with a board to give a smoother surface.

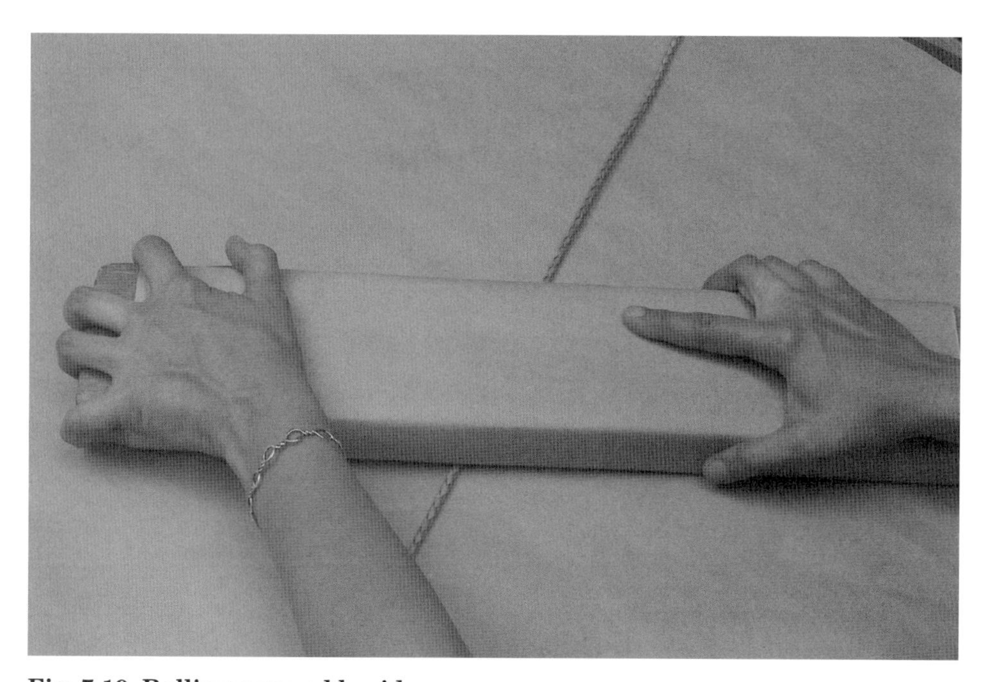

Fig. 7-19. Rolling a round braid

The braid is placed on a firm flat board or surface and rolled with a short length of wood (fig. 7-19). Both the surface and the rolling board are best covered with clean paper to absorb grease on the braid and so reduce slipping.

CHAPTER EIGHT

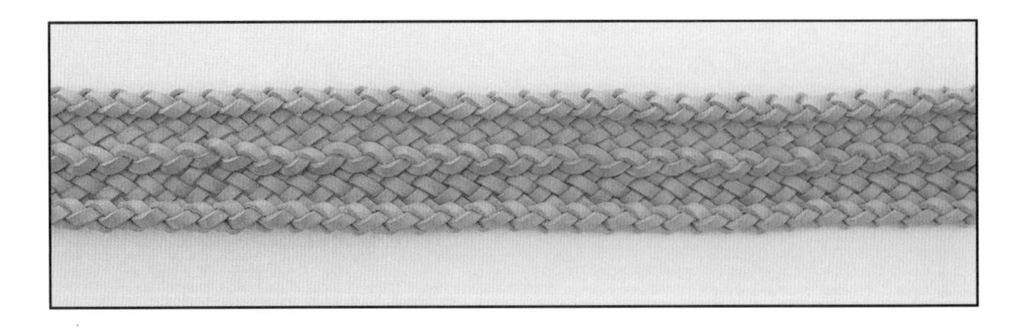

Flat Braiding

PATTERNS OF FLAT BRAIDS

The simple flat braids that use a lightweight leather such as kangaroo are made with each strand alternately going over one strand and under one in sequence. More or fewer strands of wider or narrower width are used to achieve the desired total width and appearance.

Fancy flat braiding may be carried out in three forms: variations of the over-one, under-one sequence; the use of ridges; or the use of lace in contrasting colors. Ridges may be used on flat braids as ornaments or as reinforcement. Center ridges are usually ornamental, although they may be used to give more weight to a braid, as may be needed in the central section of a braided gun sling. Edge ridges may be ornamental, or they may be used to strengthen an edge to prevent it from rolling over in use. Two-tone work is attractive in flat work. A wide variety of patterns are possible in simple flat braiding with only two colors of lace. Two-tone work using other braids or combined with ridges permits a further array of patterns. For more information on various fancy flat braids and ridge braids, see Bruce Grant's *Encyclopedia of Rawhide and Leather Braiding.*

MECHANICS OF FLAT BRAIDING

Flat braids may be started on the hook, from a yoke, or from a round braid. For belts and other usual strap work, the braid is normally started on the hook. Starting on a yoke can present attractive design opportunities for some projects. Flat work incorporated into round work is found in such items as dog leads with flat wrist loops and a bolo tie with a flat section over the neck.

Flat Braiding in Four Strands

A flat braid in four strands is started on a hook by placing one middled strand on the hook and interweaving the other strand just below the hook. Place the middled free strand over the right-hand strand hanging from the hook and under the left strand. Cross the two strands hanging from the hook and bring the braid snugly up to the hook as shown in figure 8-1. To start a flat braid in four strands from a yoke, cross the two center strands right over left. Put the upper right-hand strand under the lower right-hand strand, and the upper left-hand strand over and under the other two strands on the left, as shown in figure 8-2.

The braiding procedure used to form a neat firm four-strand braid is shown in figure 8-3. The upper strand on the right is woven under the lower strand, twisting it closely into place. The upper strand on the left is then woven over the adjacent strand and under the next, again being put closely in place. Both worked strands are then tightened by pulling all four strands sideways at a flat angle.

Flat Braiding in Six Strands

A flat braid in six strands is started on the hook by placing one middled strand on the hook and interweaving the other two middled strands. Place the two free strands over the right-hand strand hanging from the hook. The upper of the two free strands is placed under the left-hand

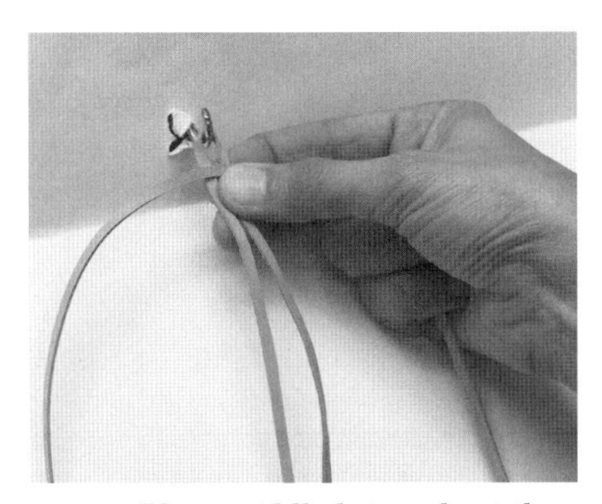

a. Place middled strands at the hook.

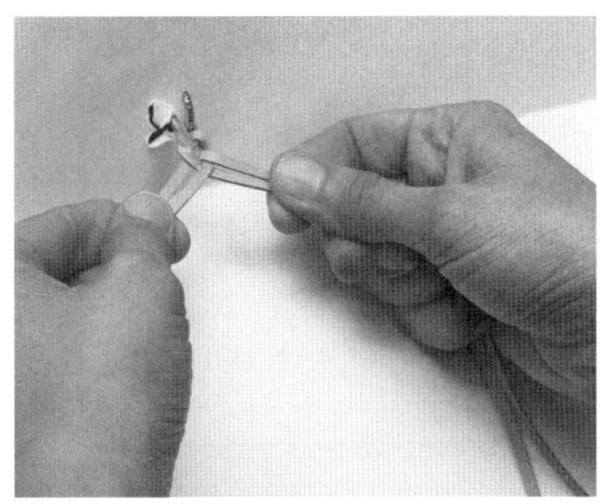

b. Start (first two strands woven to start the braid).

Fig. 8-1. Flat braid in four strands started on a hook

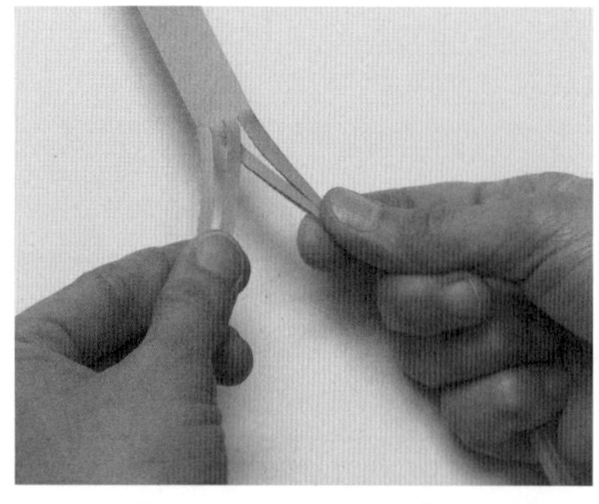

a. Cross center strands at yoke.

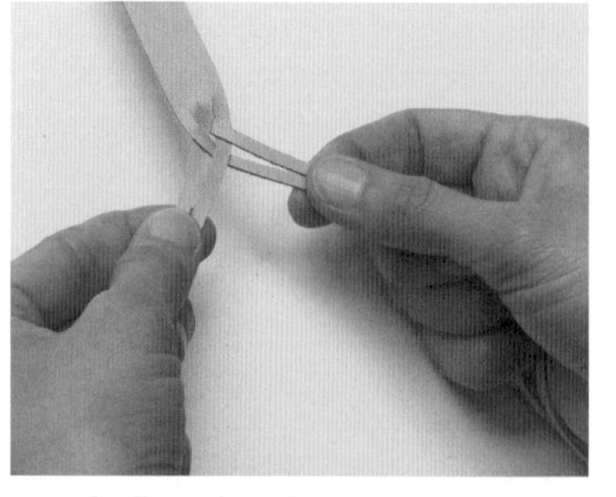

b. Start braiding.

Fig. 8-2. Flat braid in four strands started from a yoke

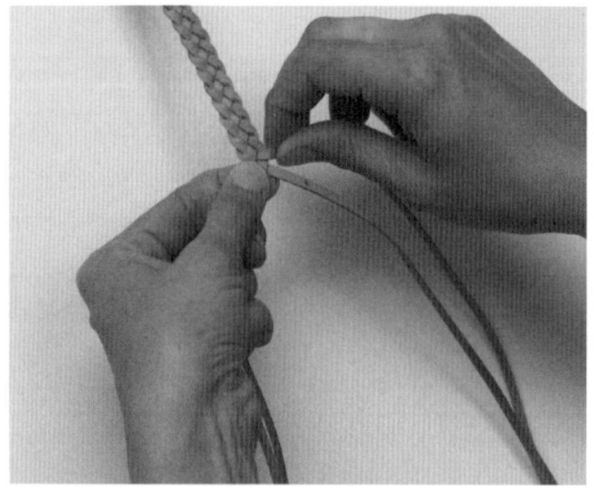

a. Weave the upper right strand under the lower strand and twist it into place.

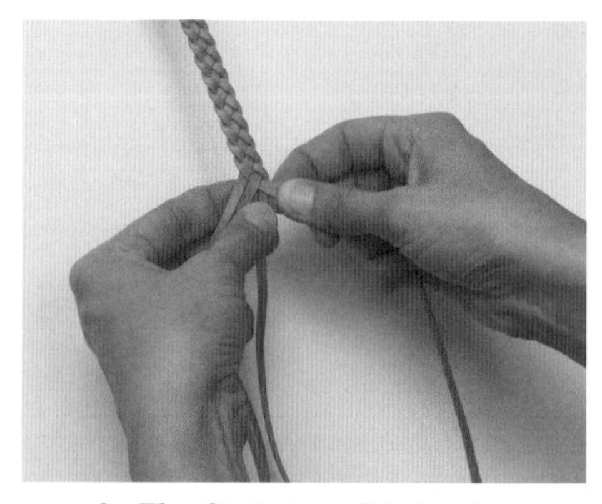

b. The first strand is in place.

c. Weave the upper left strand over the adjacent strand and under the next.

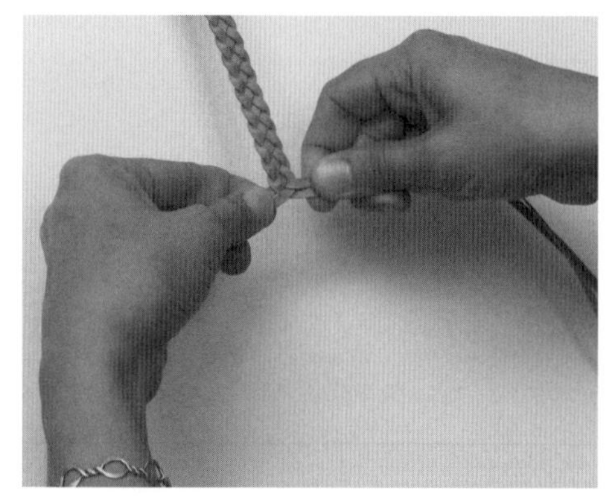

d. Tighten the braid by pulling all four strands sideways.

Fig. 8-3. Braiding procedure for four-strand flat braid

strand hanging from the hook and the lower free strand is placed over this strand. The three strands on the left are now held by the left hand; the second strand from the top is woven under the strand hanging from the hook and the two strands hanging from the hook are crossed left over right (fig. 8-4). The entire assembly is brought close together with no gaps. In starting from a yoke, the two center strands are crossed left over right. The outer strand on the right is then woven under the adjacent strand and over the next, and the outer strand on the left similarly woven over, under, and over the adjacent strands (fig. 8-5).

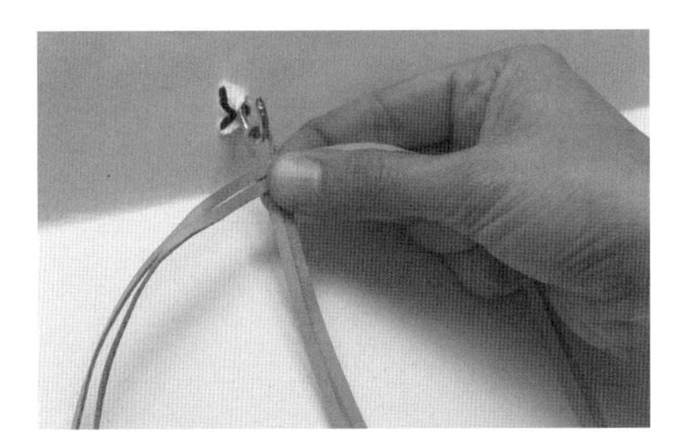

a. Place one middled strand on the hook and hold the other two middled strands on top of this doubled strand just below the hook.

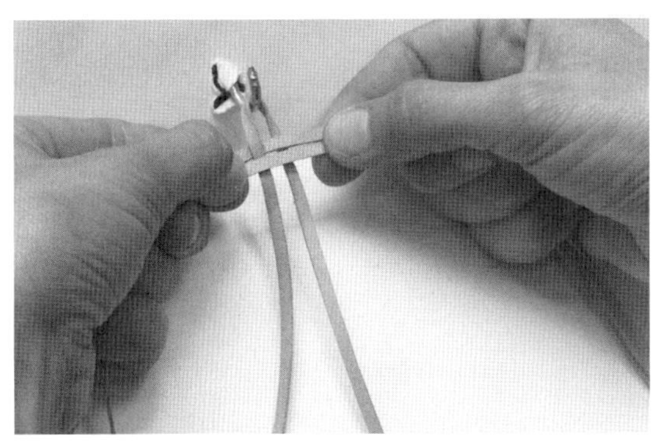

b. Weave the left-hand strand from the hook over the first of the two strands and under the second.

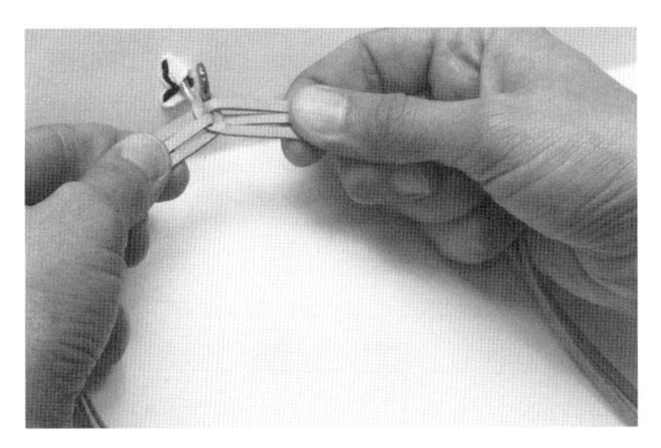

c. Weave the right-hand strand from the hook under the first of the two strands and over the second. Then, bringing the left-hand strand from the hook to the right, weave the right-hand strand from the hook under it.

Fig. 8-4. Starting a six-strand flat braid on the hook

a. Cross center strands at yoke, left over right.

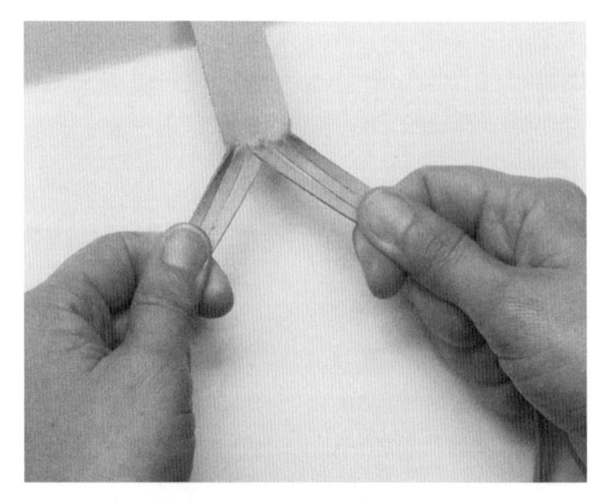

b. Weave the upper right strand under the adjacent strand and over the next.

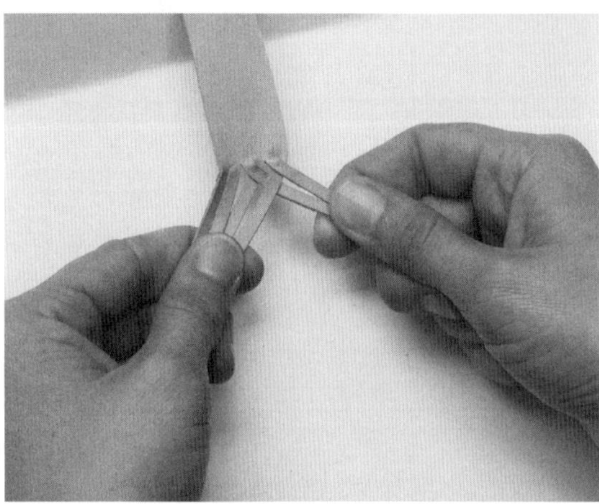

c. Weave the upper left strand over the adjacent strand, under the next, and over the last.

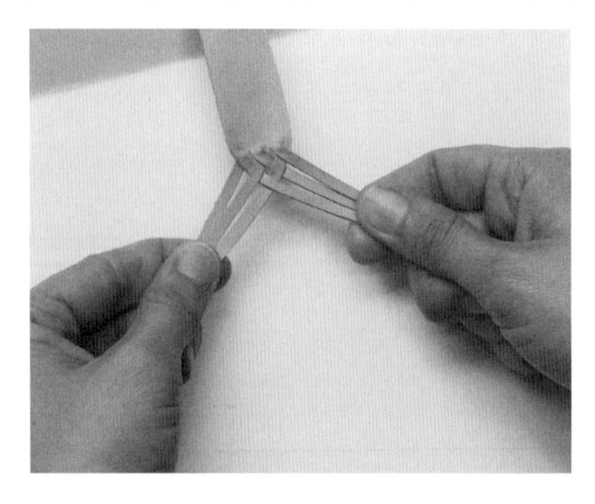

Fig. 8-5. Starting a six-strand flat braid at the yoke

As with the four-strand, the braiding procedure to form a neat firm braid is to interweave one strand from each side closely into place, then tighten both worked strands by pulling all six strands sideways at a flat angle, as shown in figure 8-6.

Flat Braiding in Eight or More Strands

A flat braid in eight or more strands is started on the hook in the same manner as that in four or six strands. One middled strand is placed on the hook, and the other three or more strands are held in the right hand and interwoven with the strand hanging from the left. The strands on the left are then held in the left hand and the hanging strand from the right is interwoven with the three strands. The two hanging strands are crossed, and the entire assembly brought closely together on the hook. The upper strand on the right is woven through, followed by the one on the left, and this process is then repeated. See figure 8-7.

The sequence for starting a flat braid of eight strands on a yoke differs from that for a six-strand braid on a yoke. The two center strands are crossed right over left, as are the adjacent pairs on each side. The outer strands are woven through in the usual fashion. It is to be noted that the center strands for the six-strand braid are crossed left over right, rather than right over left as for the eight. The purpose of this difference is to set up the braid so that in both cases the braid will be carried out with the strand worked from the right going under the first strand, and the strand worked from the left going over the first strand. In this way the braiding is kept to the same style—the right-hand working strand always starts under the next strand, and the left-hand working strand always starts over the next strand, rather than jumping back

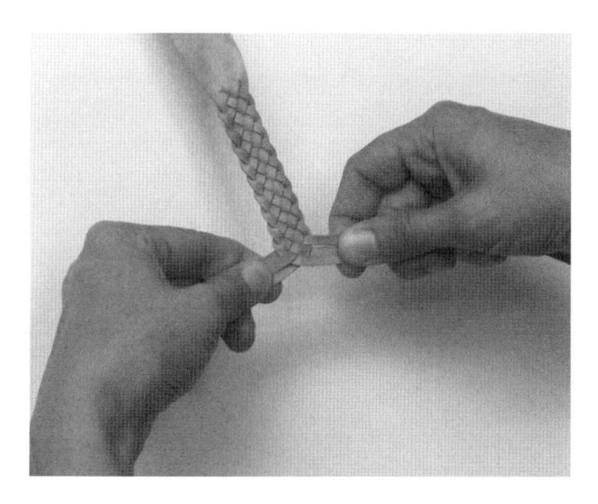

Fig. 8-6. Tightening a six-strand braid

and forth when braids of different numbers of strands are made. As a general rule in starting on a yoke, the center strands should cross right over left if the number of strands is divisible by four. If the number of strands is divisible by two, but not by four, the center strands should cross left over right. Braids with an uneven number of strands do not fit this rule. Apart from the three-strand, such braids are usually fancy braids with their own requirements. The start of an eight-strand flat braid on a yoke is shown in figure 8-8.

The procedure for braiding with eight or more strands differs from the method used with four or six strands in the way the braid is kept tight. The strand to be worked is first pulled to tighten it back to the opposite side of the braid (figures 8-9a and 8-10a). It is then closely woven through the strands and pulled firmly into position (figures 8-9b

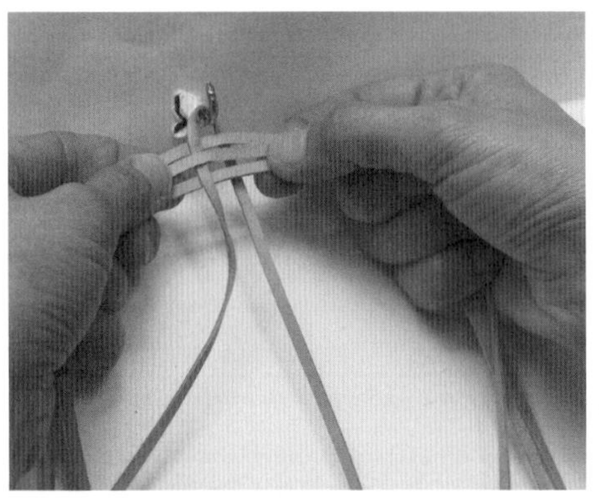

a. Place one middled strand on the hook and interweave strands on the left.

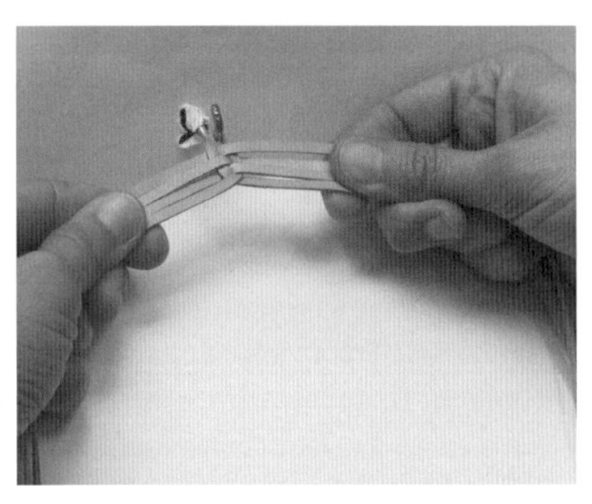

b. Start the braid.

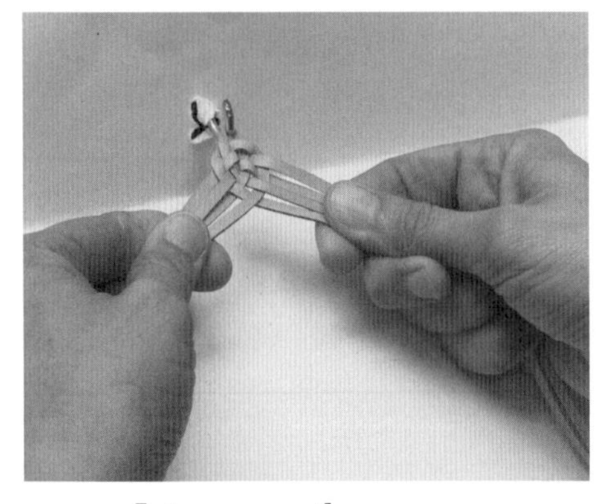

c. Interweave the upper strands.

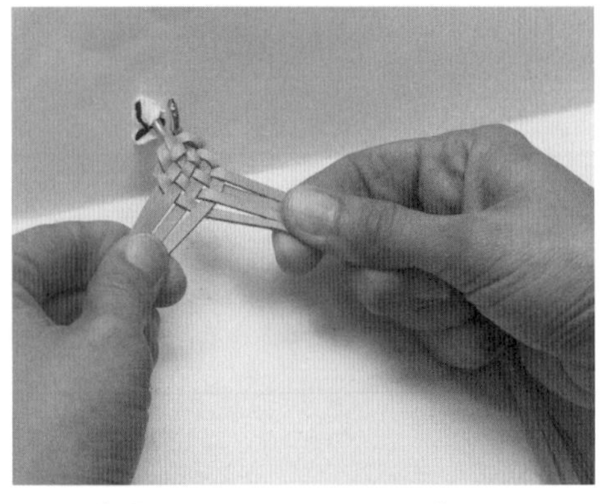

d. Interweave a second pair, showing the pattern.

Fig. 8-7. Starting an eight-strand flat braid on the hook

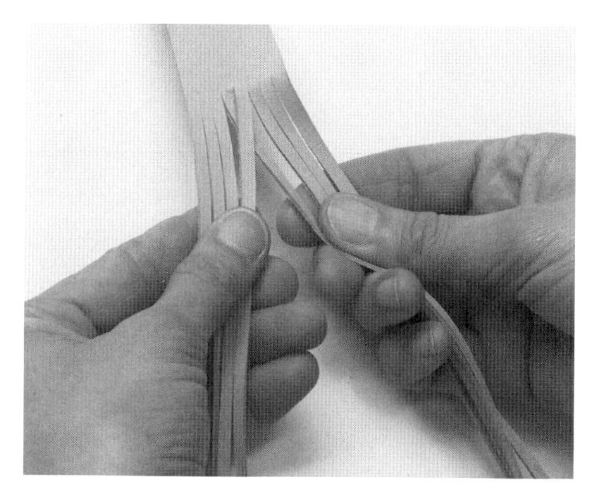

a. Cross the center strands at the yoke, right over left.

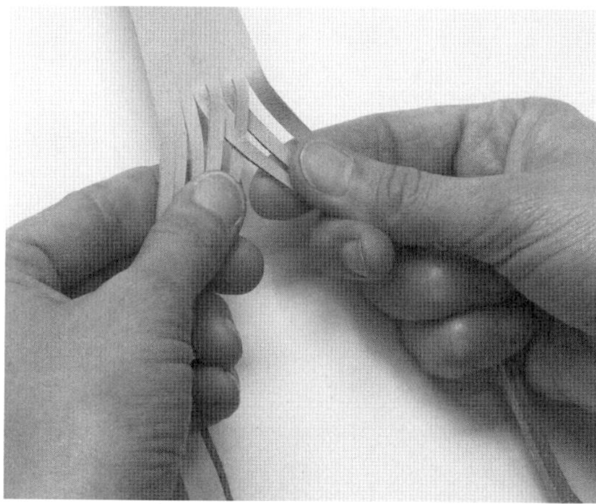

b. Cross the adjacent pair on the right, right over left, and weave the top strand of this pair through.

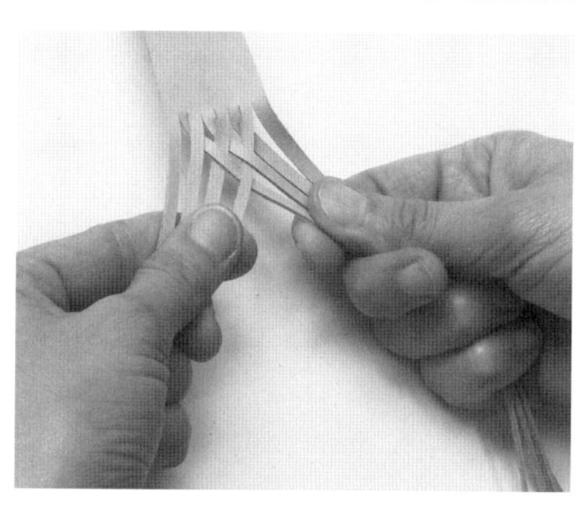

c. Cross the pair on the left adjacent to the center pair, right over left, and weave the bottom strand of this pair through.

Fig. 8-8. Starting an eight-strand flat braid on a yoke

a. Step one: Pull the working (upper right) strand tight before weaving.

b. Step two: Pull the working (now lower left) strand tight after weaving.

Fig. 8-9. Keeping the braid tight, right hand

and 8-10b. To help retain symmetry, two strands can be worked from each side, the first making the number of strands even on the two sides, the second leaving two strands more on the opposite side.

POINTS TO WATCH REGARDING FLAT BRAIDING

Good quality flat braiding requires uniformity throughout, both in the materials used and in the process of braiding. The following points should be noted.

Evenness of the Braiding

Good quality work is recognized by its evenness, tightness, and if stressed in use (as are belts), by its resistance to the development of unevenness during use. The lace used in flat braids must be firm and even. Lace that is too stretchy or uneven in properties cannot be braided well, nor will it retain an even appearance after use. Lace used in braids that will be under stress should be firmly stretched before braiding.

Paring Lace for Flat Work

The strands may be pared or not, as desired. Strands not pared produce a more distinct texture in the finished braid. If the strands are pared, the two flesh corners are pared, leaving the grain to show on the finished side of the work.

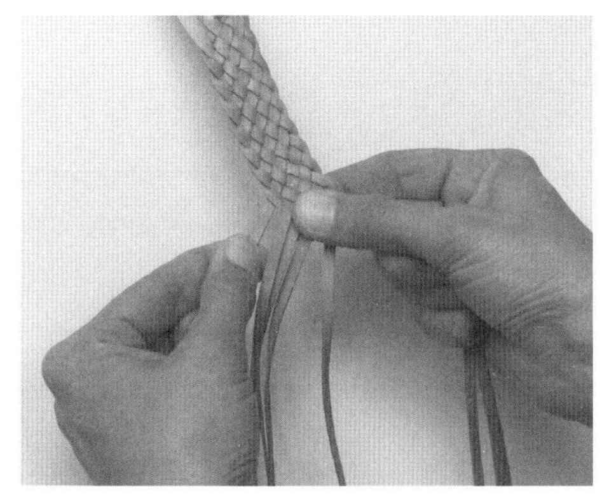

a. Step one: Pull the working (upper left) strand tight before weaving.

b. Step two: Pull the working (now lower right) strand tight after weaving.

Fig. 8-10. Keeping the braid tight, left hand

Lubrication

Braiding soap is helpful in achieving a tight braid. For hatbands (where grease might be objectionable) or where the band is neither braided tightly nor stressed in use, talcum powder is a satisfactory lubricant.

Rolling Flat Braids

After braiding, flat braids may be rolled or not as desired. If rolled, a cylindrical roller such as a rolling pin may be used. Rolling makes a smoother surface but does little to improve poor work.

Fig. 8-11. A two-tone twelve-strand belt and a sixteen-strand belt with dees

CHAPTER NINE

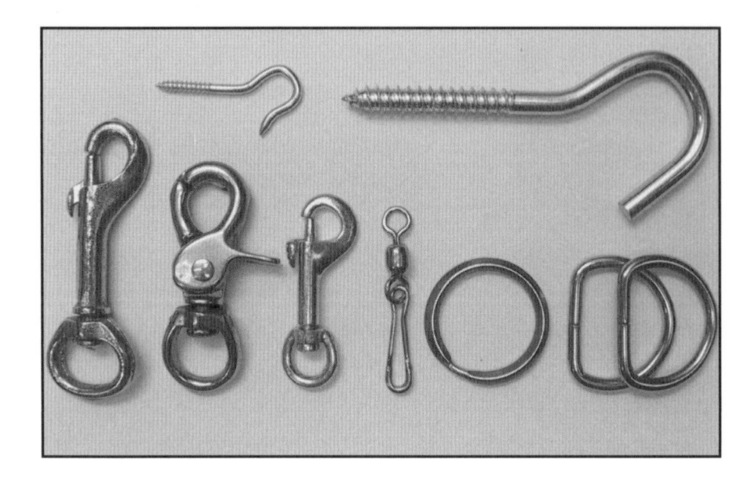

Getting Started on Projects

The projects shown here are designed to demonstrate the application of the basic techniques and to present some secondary techniques used in making specific items. Insofar as possible, items have been selected to present one technique or concept at a time. The projects are not vehicles to learn the basic techniques, but rather examples to demonstrate the use of good technique and design. As such, the basic techniques used in any project should be learned and practiced before the project is started. For example, before starting project 1, the key lanyard, the necessary techniques of cutting lace (if starting with a kangaroo skin), paring, applying braiding soap, and braiding a four-strand round braid should be learned and a reasonable facility in using them developed. The effects of varying the width and thickness of lace, the angle of paring, the amount of braiding soap, and the tightness of the braid should all be understood by practicing with test pieces. When these factors are under control there is every prospect that the key lanyard will turn out well and that the time spent on this project and the others will be both productive and satisfying as a learning experience.

When cutting lace for these projects from a kangaroo skin, the leather for cores and lace for the simple four-strand round braids may be taken from the outer part of the skin. The lace for belts should be taken only from the very best part of the skin. The quality of leather used for dog leads and hatbands is less critical; these items do not require the firm material needed for belts. The projects are presented to follow the order that the lace may be cut from the skin, with the exception that the core for the dog lead should be either cut ahead of time from the outer part of the skin or taken from some other thin leather.

All projects can be done with precut lace. The lace should be sorted out before starting the projects so that the firmest section will be used for the belt, the most stretchy in the lanyards, and the in-between for the dog lead and the hatbands. Cores can be cut from any thin leather, or one or more thin strands of lace may be used.

The projects are designed to be carried out in order, and the secondary techniques presented in one project may be used in later projects. If the order of undertaking the projects is changed, or any of the projects are omitted, it may be necessary to refer back to earlier projects for a more detailed presentation of some techniques.

Chap.	*Project*	*Techniques and Components Introduced*	*Techniques Used*	*Materials Needed*
10	Project 1: Key lanyard, four-strand round braid	Back-braiding Lark's head knot Back-braiding round to round	Four-strand round Paring	One key ring Two 70-inch lengths of ⅛-inch lace
11	Project 2: Whistle lanyard, four-strand round braid	Tassel with covering knot Sliding knot	Four-strand round Paring	Swivel snap Waxed string Two 112-inch lengths of ⅛-inch lace Four additional strands of ⅛-inch lace: two 8-inch, one 12-inch, and one 14-inch
12	Project 3: Key lanyard, six-strand round braid	Starting from a yoke Braiding in single-diamond six strand Changing between four-seam and single-diamond braiding and back; changing from diamond to flat braiding Back-braiding flat to round	Four-seam six-strand Six-strand flat Covering knot Paring	Key ring Yoke with three 35-inch strands on one side and two 35-inch strands and a 47-inch strand on the other; or two 70-inch and one 84-inch lengths of ⅛-inch lace

Chap.	*Project*	*Techniques and Components Introduced*	*Techniques Used*	*Materials Needed*
13	Project 4: Dog lead, eight-strand round braid	Braiding with a heavy core	Four-seam eight-strand Eight-strand single-diamond Eight-strand flat Back-braiding flat to round Paring	Bolt snap Core 27 inches by 3/8-inch to 1/4-inch wide Yoke with four 1/8-inch strands 42 inches long on each side; or two 1/8-inch strands 84 inches long
14	Project 5: Hatband, four-strand flat braid	Fringe from split yoke Double-half-hitch tassels	Four-strand flat Covering knots as keepers	Four-inch yoke with four 1/8-inch wide strands 42 inches long on one side, or two 1/8-inch strands 84 inches long Two 1/8-inch wide laces each 16 inches long
15	Project 6: Hatband, ten-strand flat braid	Keepers from single strands	Ten-strand flat Double-half-hitch tassel	Five 84-inch lengths of 1/8-inch wide lace Two 1/8-inch wide laces, each 8 inches long
16	Project 7: Belt, twelve-strand flat braid	Turning strands over for belts Back-braiding twelve-strand flat braids Braided keepers	Twelve-strand flat	Two 1-inch harness dees Six 112-inch lengths of 1/8-inch lace for a 34-inch waist size, longer for larger waist sizes Four laces 1/8 inch wide, 16 inches long

CHAPTER TEN

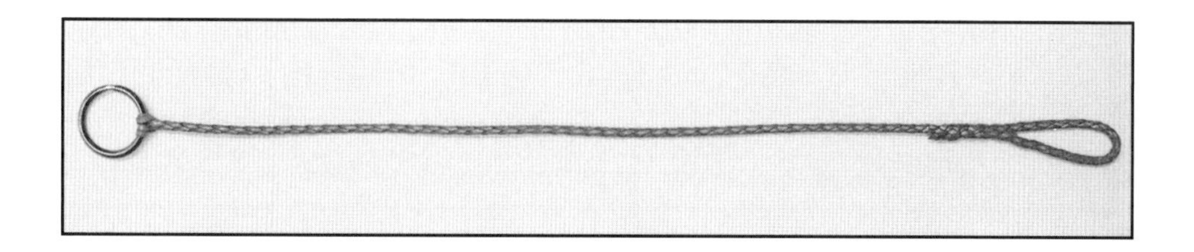

Project 1: Key Lanyard, Four-Strand Round Braid

Key lanyards may be made in several styles. A basic design is illustrated—a round four-strand braid without a core. The overall length of the lanyard, exclusive of the ring, is 19 inches. The length of the belt loop is 2½ inches. The material is ⅛-inch-wide kangaroo lace.

1. Cut two strands of ⅛-inch lace about 70 inches long. Middle the two strands and put them on the hook with a couple of hitches. Test the strands for hidden weaknesses by pulling them. It is much better to break a strand at this stage than later. Pare the two flesh corners of each strand (fig. 10-1). It is not necessary to pare right up to the hook.

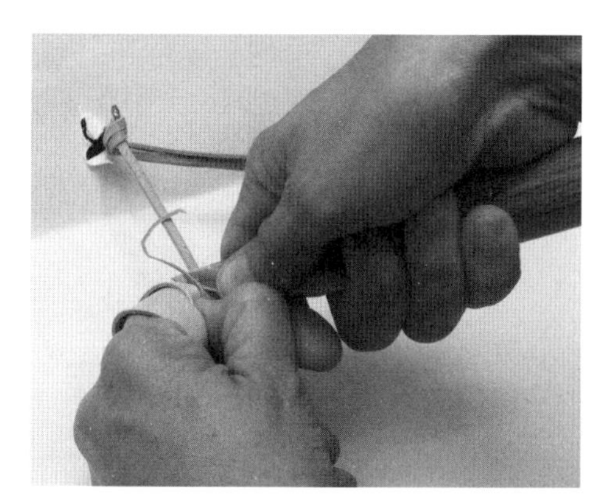

a. Pare left corner.

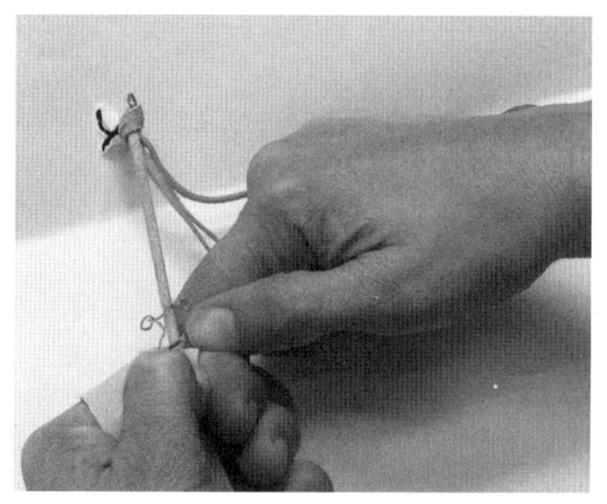

b. Pare right corner.

Fig. 10-1. Paring the corners of the lace

2. Form a lark's-head knot on the key ring at the middle of the two strands (fig. 10-2). To do this, place the middle sections of both strands together and form a loop. Put the loop through the ring. Bring the loop over the ring and pull it down to form a lark's-head knot. Adjust the knot so that one strand is neatly on top of the other and the grain sides are up. Pull tight.

a. Put loop through key ring.

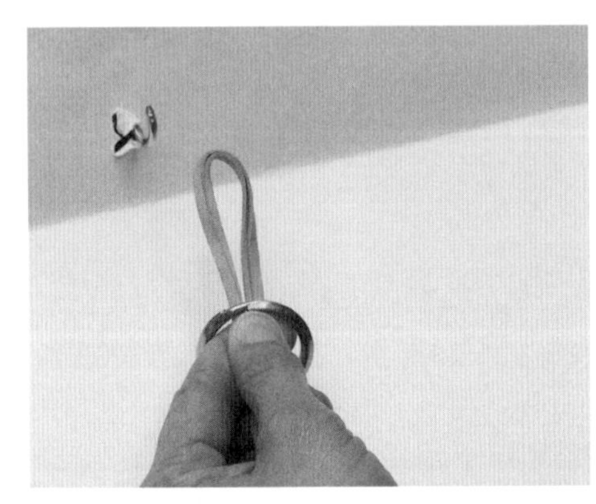

b. Put loop over key ring.

c. Pull to form lark's head.

Fig. 10-2. Lark's-head knot on the key ring

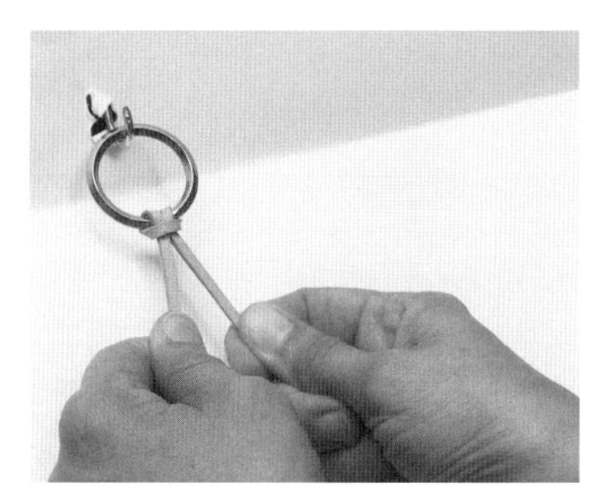

d. Adjust the strands so that one is neatly on top of the other with the grain side up, and pull tight.

3. Place the ring on the hook and apply a coating of braiding soap to the strands, coating both sides evenly. To begin braiding, spread the four strands out, the two strands on top being crossed with the right-hand strand over the left. The first stitch takes the upper right strand around the back and between the left-hand strands, to be laid in place in front. The second takes the upper left strand around as in a normal four-strand round braid (fig. 10-3).

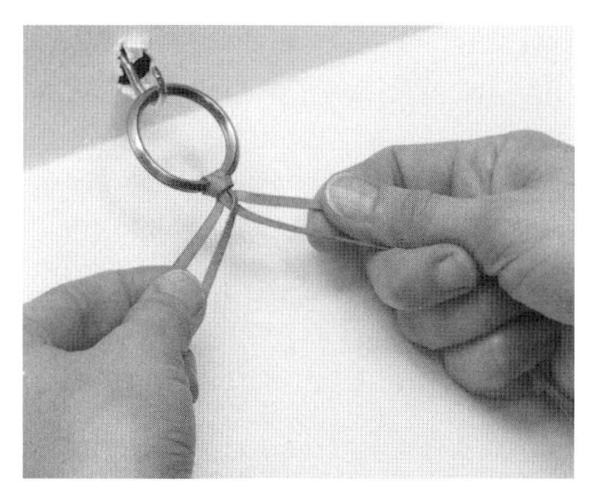

a. To begin, cross top strands, right over left.

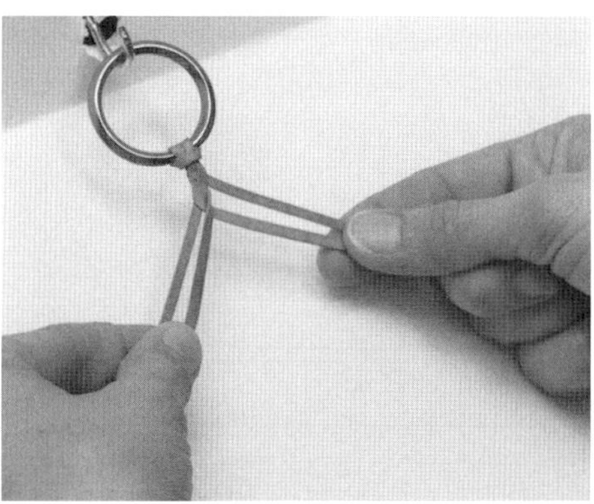

b. Work the other two strands in the normal way to start the braid.

Fig. 10-3. Braiding the lanyard

4. Continue braiding to a length of 23½ inches, maintaining a uniform pull (fig. 10-4). The additional two inches beyond that of the finished product allow for any loosening of the last few stitches or shrinkage during rolling. Check that the braid is firm and solid and that the seams are straight. A braid spiraling one way or the other usually indicates one hand is pulling harder than the other.

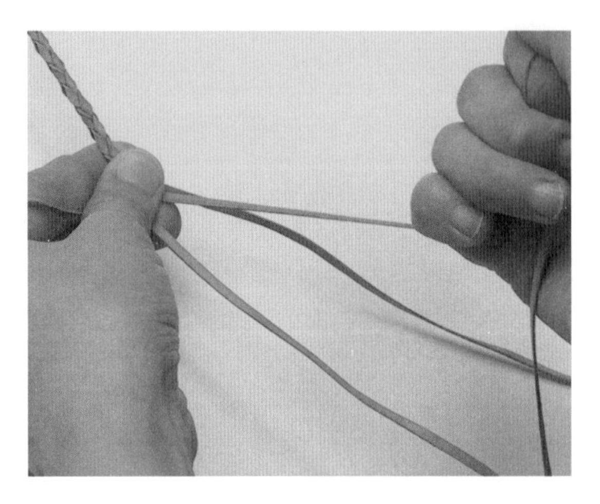

Fig. 10-4. Continue braiding. Use a firm even pull.

5. Pull all four strands to tighten the last section of braid (fig. 10-5).

6. Tie off the braid with a slip hitch, using either of the upper strands (fig. 10-6). The remaining length of the strands should be noted. If the lace is uniform, the ends should be the same length on each side.

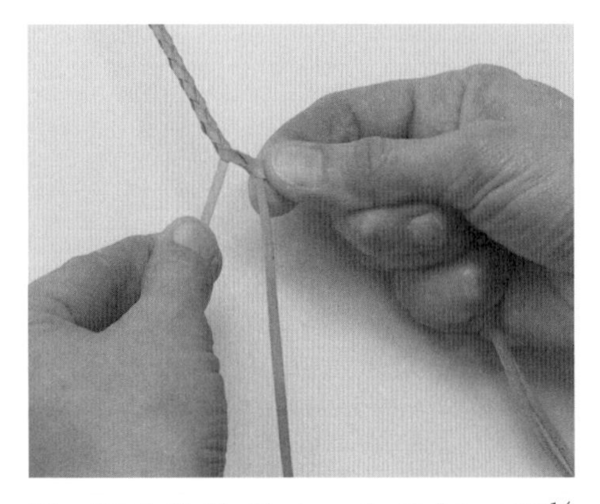

Fig. 10-5. Pull all strands tight at 23½ inches.

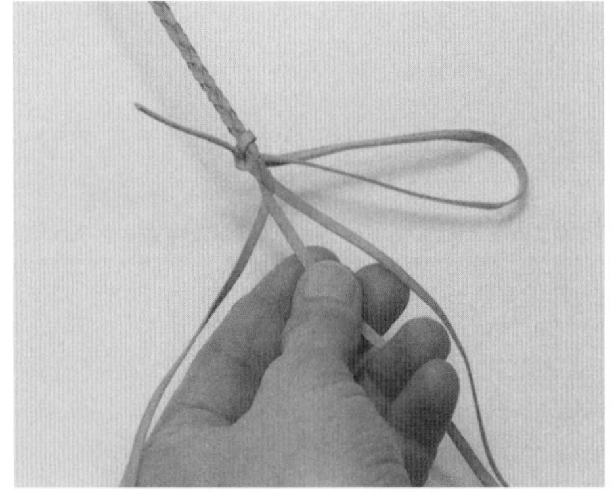

Fig. 10-6. Tie off with a slip hitch, ready for rolling.

7. Roll the braid. The ring should be placed over the edge of a table or rolling surface, so that it does not interfere with the movement of the braid during rolling (fig. 10-7a).

Rolling makes the surface of the braid smooth and round. It will not eliminate the effect of imperfections such as loose or excessively tight stitches, although it may make them a little less obvious. A good firm braid rolls evenly and smoothly under fairly heavy rolling pressure. An excessively tight braid rolls hard, forming distinct corners. A loose hollow braid feels soft and squishy during rolling. It may be rolled gently, to give a braid that initially looks acceptable. However, when used, it will soon show up as loose and sloppy. The only cure for a loose braid is to undo it and braid it again more tightly.

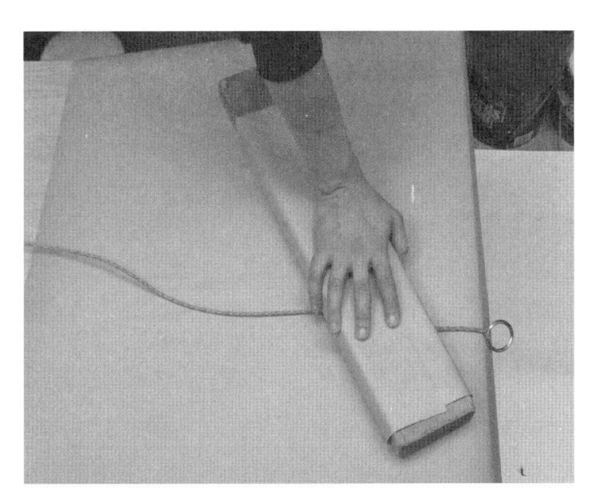

a. Roll the braid with the ring over the edge of the rolling surface.

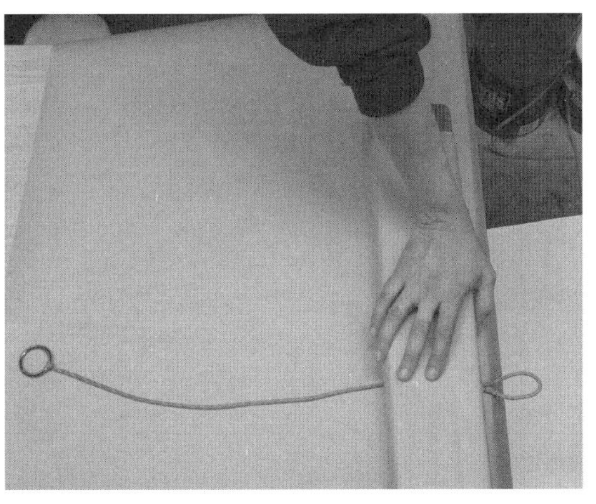

b. Roll the back-braided portion with the loop over the edge of the rolling surface.

Fig. 10-7. Rolling the braid to achieve a smooth round surface

8. To form the belt loop, first cut points on the ends of the strands. Long points, about ½ inch long, are easiest to work. Stretch the braid and measure out 21½ inches. Undo any extra braiding back to this length, and form a loop 2½ inches long. The four free strands are positioned just over the four strands in the main braid. The free ends will be woven through so that one end lies on top of each of the four strands. Follow the sequence in figure 10-8. Note that a lacing needle has been placed on the loop to show which side of the loop is pictured.

Start with a free end that is found underneath the place where the free ends were last crossed in the braided end. Using a fid, put the free end under the strand in the main braid directly under the place where it was last crossed in the braided end. The adjacent free end will go over this strand and under the next. These two strands thus take up pathways on top of adjacent strands. Turn the loop over and do the same with the remaining two free strands in the other direction. Next weave all free ends through on top of the strands they are following until the one nearest the loop has passed under three strands. Weave the free ends through one at a time, working the one nearest the loop each time, to minimize the times it is necessary to put an end under a doubled strand. When the end nearest the loop has passed under three strands, continue weaving the other ends until they are on the same side of the braid. The free ends are then cut off, leaving about ¼ inch excess, and the back-braided section is firmly rolled (fig. 10-7b).

a. Insert first end (side one of loop).

Fig. 10-8. Forming the belt loop

b. Insert second end (side one of loop).

c. Insert third end (side two of loop).

d. Insert fourth end (side two of loop).

Fig. 10-8. *Continued*

e. Bring free ends to one side.

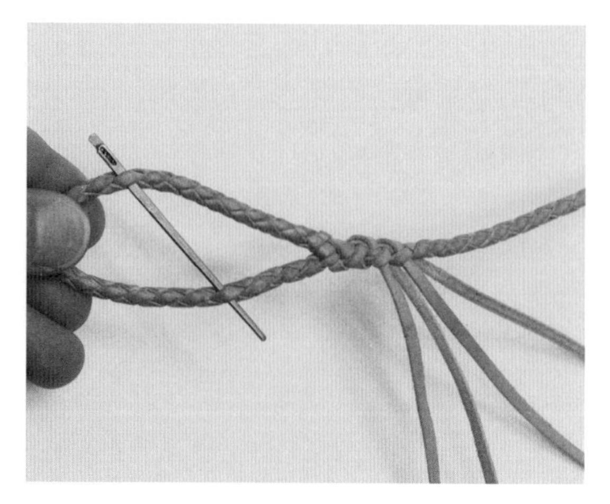

f. Cut off ends.

The lanyard should be finished with a light coat of shellac. The shellac should be cut from full strength with a little thinner (about 10 percent). It should be lightly brushed on using the end of the brush with multiple strokes, with the brush carrying only a little shellac. Do not flow shellac on with a loaded brush.

FURTHER SUGGESTIONS

A lightweight dog lead can be made in a similar fashion using lace that is either ⅛ inch wide or ¼ inch wide. Use a small bolt snap. An overall length of 4 feet with a loop of 6½ inches will be satisfactory, requiring two strands, each approximately 13½ feet long, to complete this project.

CHAPTER ELEVEN

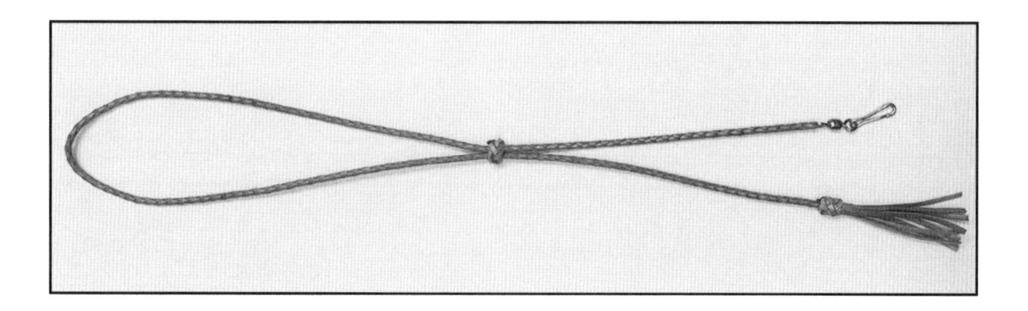

Project 2: Whistle Lanyard, Four-Strand Round Braid

A whistle lanyard may be made from a four-strand round braid. The finished length is 36 inches, plus swivel snap and fringed tassel. As with any hardware used with braiding, the swivel snap should be of good quality stainless steel or brass. Swivels and snaps are available in a range of sizes at fishing supply stores.

Cut two strands ⅛ inch wide and 112 inches long. Tie the two strands to the hook at the middle with a lark's head, pull to stretch and test the strands, pare both flesh corners, and apply braiding soap. Do not pare the lace in the middle where it is tied to the hook, as this will leave more leather where it is attached to the swivel.

Put the two strands through the swivel eye and middle them, grain-side uppermost. The portion of the lace at the middle being not pared will provide enough body to take normal wear if the braid at the swivel is tight.

1. To start the four-strand braid, take the two laces from the same middled strand and cross these on a swivel right over left (fig. 11-1a). Bring the other strand from the right around the back and up between the two strands on the left (fig. 11-1b), and the strand from the left around and up on the right to give the start of the braid (fig. 11-1c).

2. Continue to braid in four-strand to 36 inches, not including the swivel. Tie the braid off with a firm half hitch, with the free end of the half hitch facing in the same direction as the other ends (fig. 11-2). Roll the braid.

a. Cross the two strands from one middled lace on the swivel, right over left.

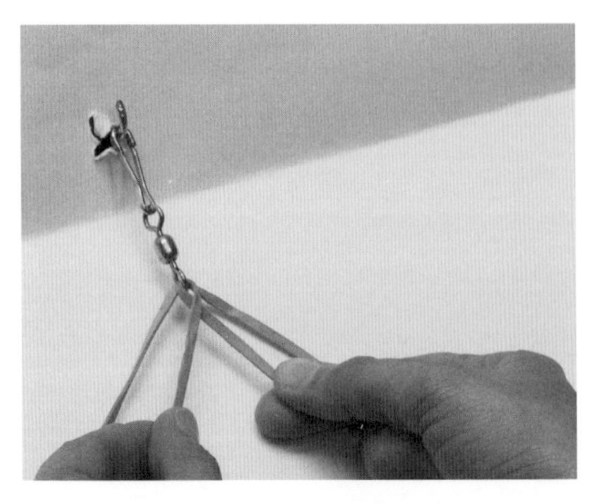

b. Work upper right strand.

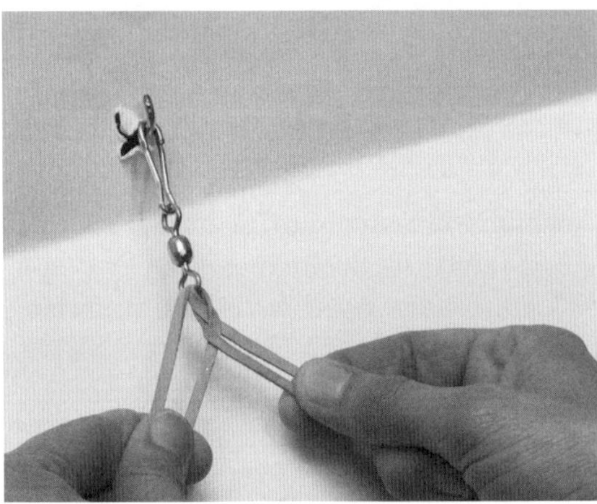

c. Work upper left strand to complete the start of the four-strand round braid.

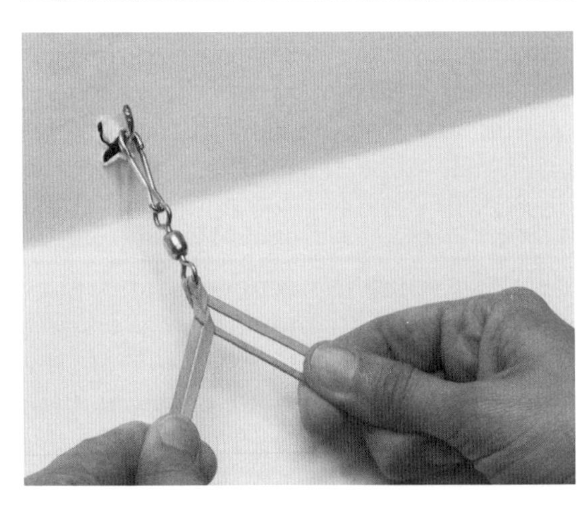

Fig. 11-1. Starting the four-strand round braid

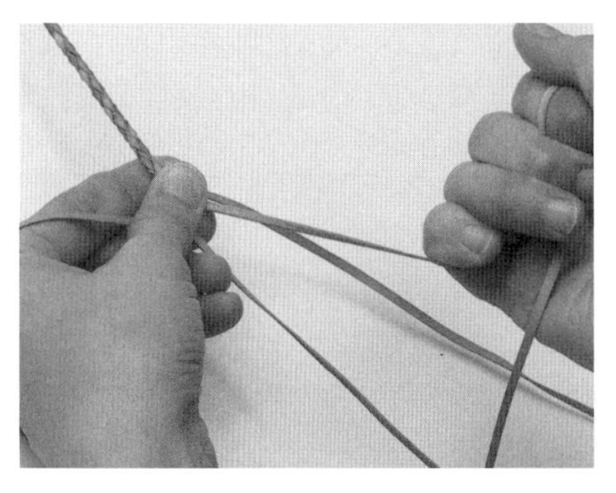

a. Continue braiding in four-strand, using a firm uniform pull.

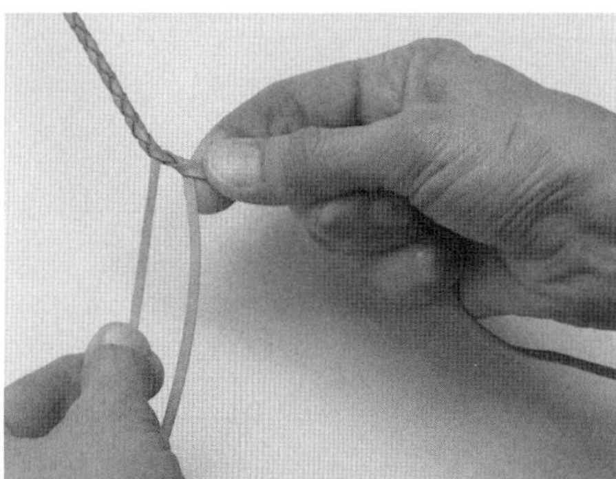

b. Pull all strands tight at 36 inches.

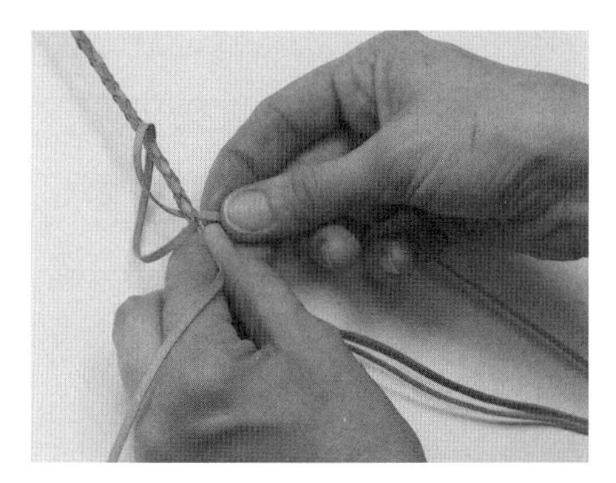

c. Form a half hitch with the free end of the half hitch facing in the same direction as the other ends.

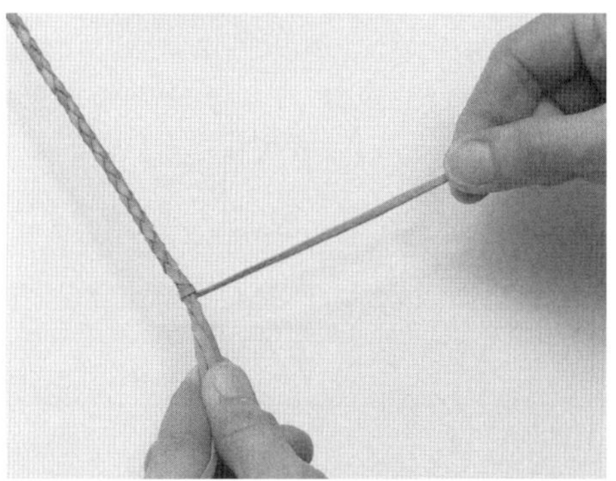

d. Pull the hitch tight.

Fig. 11-2. Finishing the braid

3. Form a tassel on the end with the half hitch. Start by cutting two ⅛-inch-wide strands 8 inches long. These will be used to add strands to the fringe. Using thin waxed string tied to the hook, put a half hitch just above the lace half hitch used to tie off the end of the braiding (fig. 11-3).

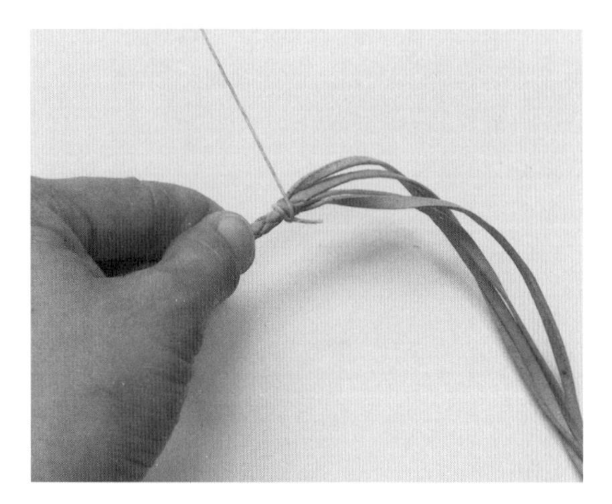

Fig. 11-3. With thin waxed string, put a half hitch just above the lace half hitch.

4. Place the two 8-inch strands, middled at the string, with their grain-sides against the braid. Put on one turn of string to hold the added strands against the lanyard. This is most easily done by putting one strand in place and holding it with the string while the second is put in place and the turn completed (fig. 11-4).

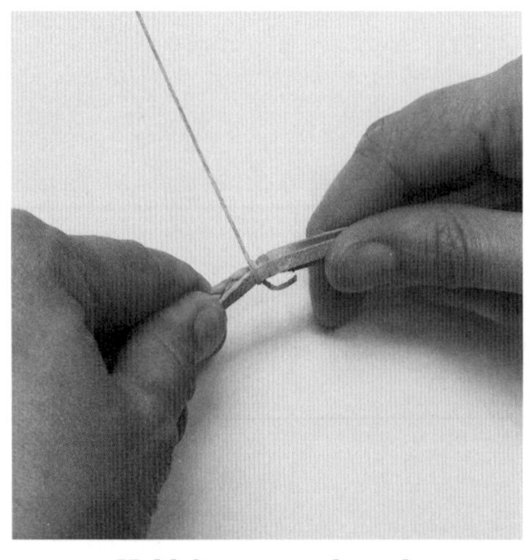

a. Hold first strand in place.

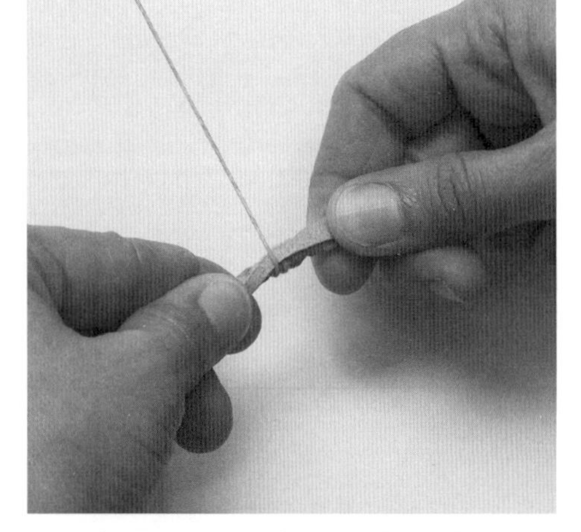

b. Put second strand in place.

Fig. 11-4. Place two strands grain-side against the braid.

5. Next, fold the upper ends of the added strands down alongside the fringe. Now put another two turns of string over the turned-down ends of the short strands (fig. 11-5). Add these extra turns just on the tassel side of the lace half hitch. Finally, tie off the string with a half hitch to secure the strands. You now have a neat basis for a covering knot.

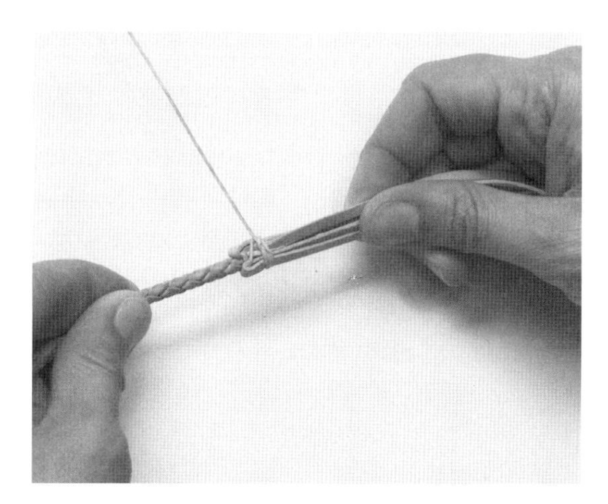

Fig. 11-5. Fold strands down alongside fringe and hold with two turns of string.

6. Put a covering knot over the tassel. This knot has a very simple structure. It consists of a series of passes, or loops, each extending the length of the knot. Each pass is rotated about the axis of the knot so that it lies beside the prior pass, going over the top of the prior pass where it crosses at the edge of the knot. Each pass goes alternately over and under each strand it crosses, so that no strand goes over two adjacent strands or under two adjacent strands in the completed knot. Follow the procedure shown in figure 11-6.

Use a piece of lace about 14 inches long, and attach a needle to its end. Start the knot from the side nearest the braid and swivel, and finish by passing the end through the knot into the fringe. To begin, lay the knot strand across the knot's foundation, allowing the end of the strand to extend a half-inch back towards the swivel. Put the first pass or loop around the braid (fig. 11-6a), extending it down over the full length of the knot foundation made from the lace half hitch, the added tassel strands, and the waxed string, then back up to cross the knot strand at its starting point. Hold this crossing point with your thumb.

Put the second pass on, rotating on the axis of the knot so that this pass lies beside the first pass. It crosses over the last half of the first pass at the fringe edge of the knot. Continuing to follow beside the first pass, bring the strand back toward the starting point for the knot, and thread the second pass under the first half of the first pass near the swivel edge of the knot (fig. 11-6b). It then passes over itself to begin the third pass.

The third pass follows the second pass. It crosses over the second pass at the tassel end of the knot, but before it does, it must go under the first pass to meet the requirement of the pattern of weaving (fig. 11-6c). After crossing the top of the second pass at the tassel edge, it goes over the first pass and under the second pass (fig. 11-6d) as it heads towards the swivel edge before going over its own start to begin the fourth pass. The fourth and final pass goes between the third and the first passes, over the first pass and under the second (fig. 11-6e) as it goes toward the tassel edge of the knot. It then crosses over the top of the third pass, and under the first pass (fig. 11-6f) as it starts back down towards the swivel edge. It then goes over the second, under the third (fig. 11-6g), and then over the fourth pass (fig. 11-6h) on its way back to the start of the knot at the swivel edge. This completes the formal knot.

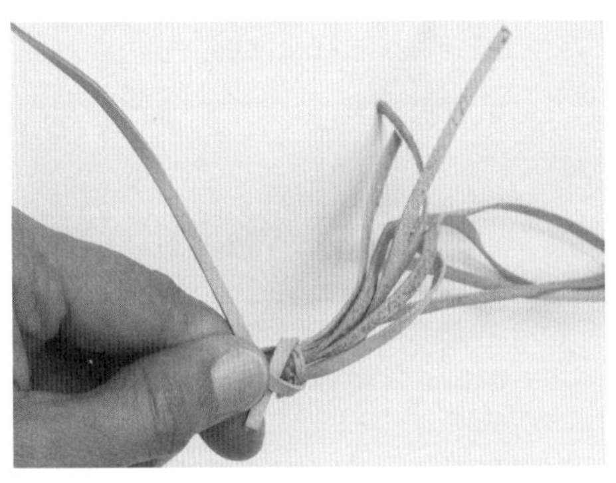

a. First pass goes around the braid.

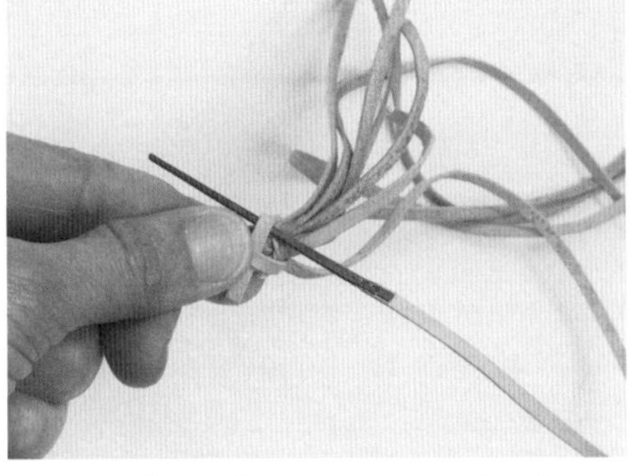

b. Second pass goes under the end of the first.

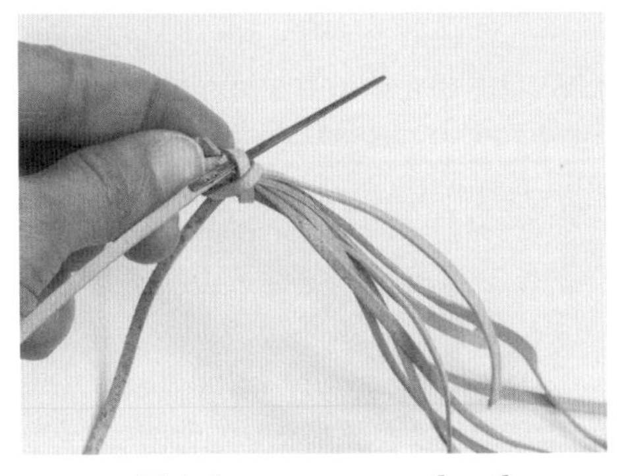

c. Third pass goes under the first pass.

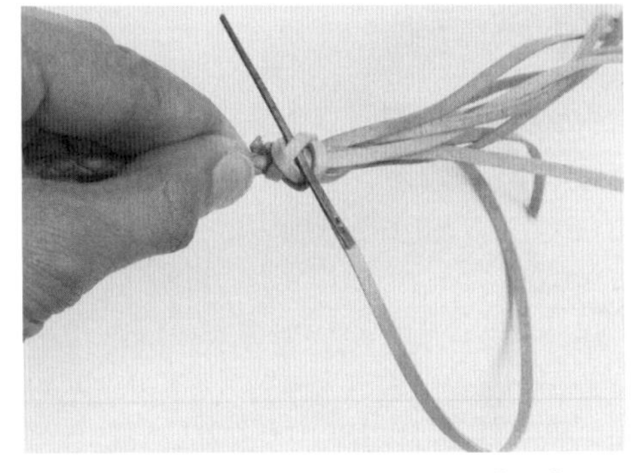

d. Third pass goes over the first pass and then under the second pass.

Fig. 11-6. Putting a covering knot over the tassel

e. Fourth pass goes over the first, under the second.

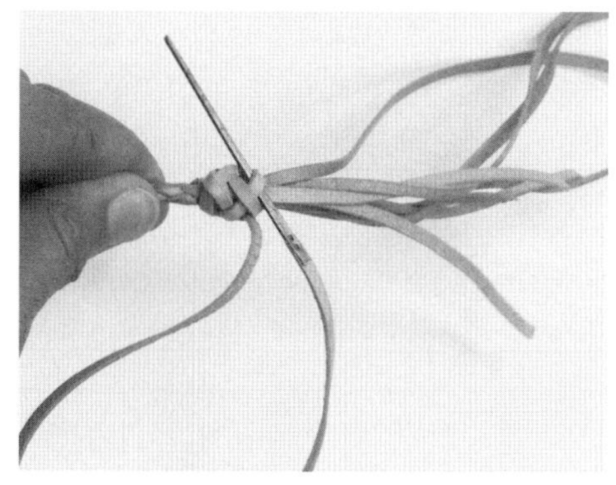

f. Fourth pass continues over the third, under the first.

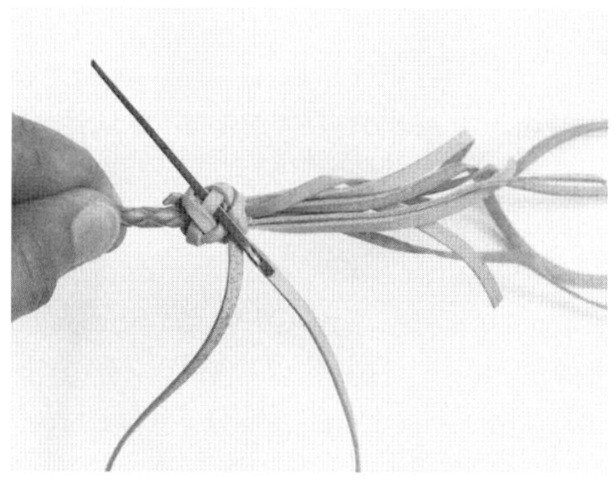

g. Fourth pass continues over the second, under the third.

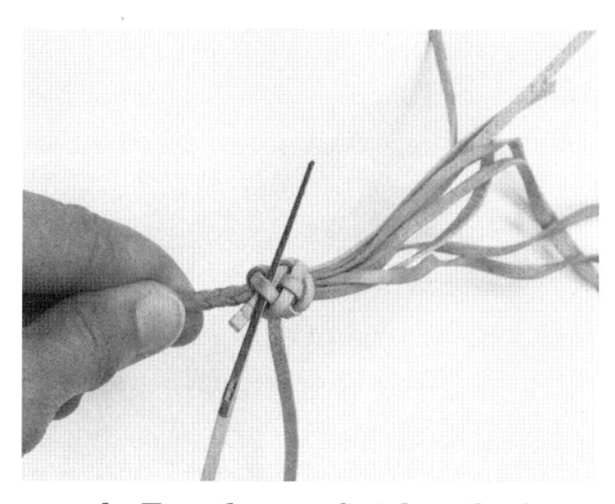

h. Fourth pass finishes the knot and goes on top of the first pass.

i. Put end into center of knot.

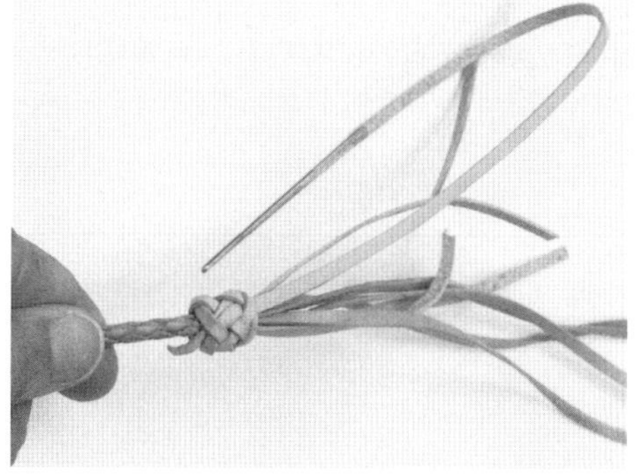

j. Tassel covering is completed.

To finish the knot in a practical fashion, the end is now worked on top of the first pass for one stitch and is then passed through into the center of the knot (fig. 11-6i) to form an additional strand in the tassel (fig. 11-6j).

Trim off any excess knot lace where the knot starts, and trim the fringe to about 2½ inches. Roll the knot lightly, and square it up with a fid.

The following point regarding knot work should be noted. The first turn should be put on snugly, but not excessively so. Each subsequent turn tightens all earlier turns, so each turn should be pulled a little tighter than the preceding turn to ensure that all turns are equally tight in the finished knot. Experience will teach the correct tension. A knot started too tight is very difficult to finish neatly. A knot started too loose must be tightened after it is finished by following around the strand with a fid, pulling the strand tight.

7. Tie on a sliding knot. The sliding knot has a very simple structure, basically an endless three-strand braid. It is made directly on the lanyard so that it will fit snugly. It is made firm by doubling or repeating the knot with a second layer which lies on top of the first. The second layer will take up some slack in the first layer, so do not make the first layer too tight. Follow the sequence in figure 11-7.

Use a piece of lace about 12 inches long on a lacing needle. To begin, fold the lanyard in half. Place the knot strand at an angle across the two braids of the lanyard, and hold the end firmly with the left thumb. With the right hand, loop the strand a little loosely around the back of the lanyard and back to the starting point, where it crosses itself to complete the first pass (fig. 11-7a). Put the second pass below the first pass, then above the first pass, and under the loose end of the first pass. The second pass now crosses over the first pass twice on its way around the knot (fig. 11-7b). This leaves two places for the third pass to be woven through. Weave the third pass over the start of the second pass, under the first pass (fig. 11-7c), over

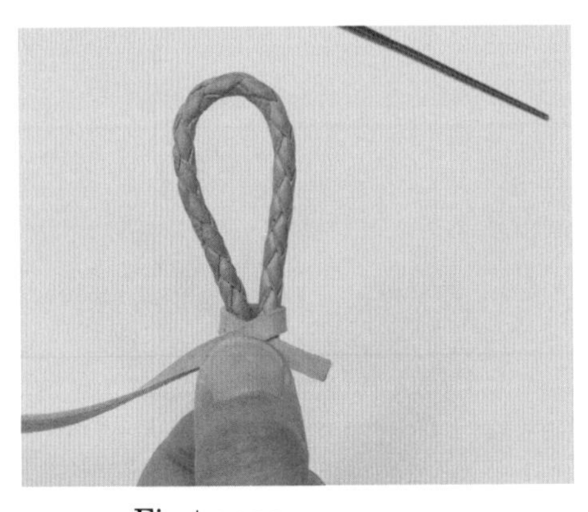

a. First pass.

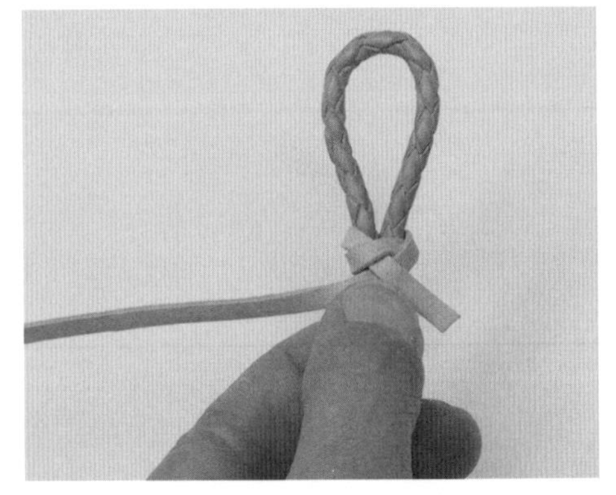

b. Second pass.

Fig. 11-7. Tying on a sliding knot

c. Third pass goes over the second and under the first (first time).

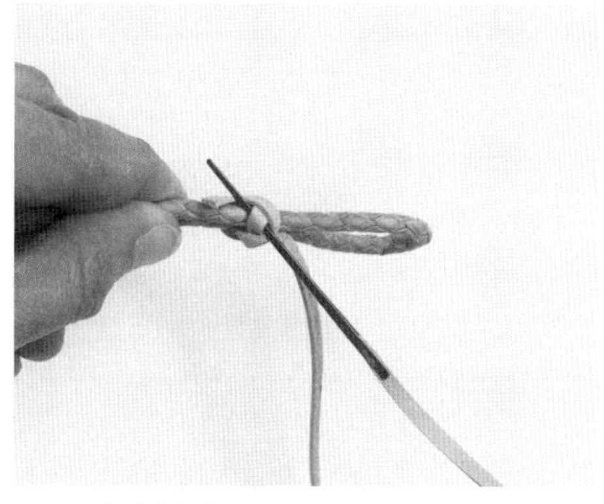

d. Third pass goes over the second and under the first (second time).

e. Start to double the knot.

f. Go over the starting end.

g. Complete the knot.

h. Trim the ends.

the later part of the second pass, and finally under the last part of the first pass (fig. 11-7d). To complete the formal knot put the lace through on top of the first pass. Now double the knot (fig. 11-7e) by retracing the path of the entire knot to create a second layer and a sturdy knot that will not come undone with use. In doubling the knot, the only irregularity is when the loose end of the first pass is met the strand is put over the loose end, rather than under. This makes a neater knot (fig. 11-7f). A fid will be necessary to make the last two or three stitches in the doubling. Finish the knot with one extra stitch, which leaves the two ends pointing in opposite directions (fig. 11-7g). Trim the ends to about ¼ inch (fig. 11-7h). Do not roll this knot, as it should remain oval to slide on the braids.

To finish, shellac the lanyard, including the knots and fringe. Do not attempt to apply any shellac under the sliding knot, as this will make the knot stick in place.

Fig. 11-8. Bolo tie

FURTHER SUGGESTIONS

Bolo ties can be made in a similar fashion. Either tie the strands on the hook to keep free ends for a fringe, or tie middled strands with a loop to provide a fringe. The standard length for a bolo tie is 38 inches plus tassels.

CHAPTER TWELVE

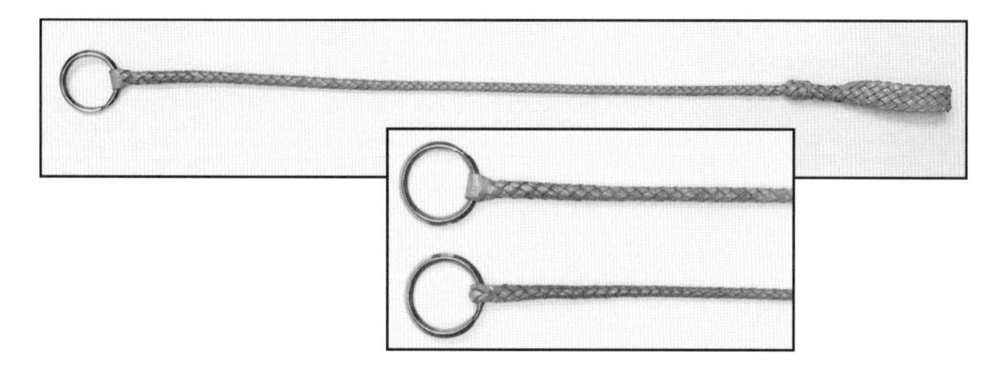

Project 3: Key Lanyard, Six-Strand Round Braid, Solid Yoke or Braided Yoke

A more ornate key lanyard may be made in a six-strand round braid. While basically similar to the four-strand lanyard, the six-strand lanyard is more attractive, particularly in having a flat loop. A simple knot is used to cover the ends of the strands where they are trimmed after back-braiding. The length of the lanyard is 19 inches excluding the ring. It is made in ⅛-inch lace. Wider lace may be used if a heavier lanyard is desired. A yoke is used where the lanyard attaches to the ring. A wear leather is placed under the yoke, to prevent wear on the yoke and to allow the ring to move more freely on the lanyard. If the project is to be made in cut lace, the yoke is made by braiding the strands into a three-strand braid to form the yoke.

1. To cut a set of strands with a central yoke for this lanyard start by cutting three strands 35 inches long, stopping them even with one another and leaving them attached to the skin. Measure ⅝ of an inch past the 35-inch mark, and start two more cuts so that three more strands can be cut leaving a ⅝-inch yoke holding them. Cut the first or top strand on this side of the yoke 47 inches long, the other two strands 35 inches long (fig. 12-1a). The longer strand will be used for making the covering knot. Tie the strands to a hook near the yoke, test them by pulling them, and pare them. The strands for a six-strand round braid may be pared on opposite corners (one flesh-side and one grain-side) or both on flesh-side corners. In this lanyard, however, since the loop is a flat braid which is best made in strands pared on both flesh-side corners, the strands should be pared on the flesh side.

To make the set from precut lace, use two strands 70 inches long and one strand 82 inches long. Middle these strands with the long strand not at its true middle but with one part 35 inches long and the other 47 inches long. Tie the strands to the hook at the middle, pull to stretch and test, and pare them. Apply braiding soap and braid a short section, about ⅝ of an inch long, at the middle (fig. 12-1b). This braided section makes the yoke.

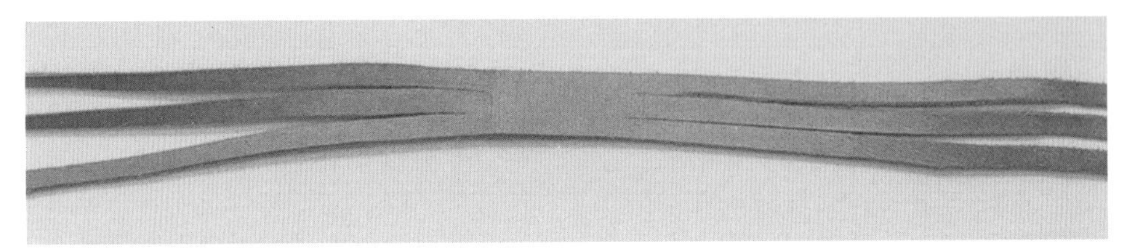

a. Solid yoke

b. Braided yoke (using precut lace)

Fig. 12-1. Sets with solid yoke and braided yoke

2. Cut a small piece of leather 4 inches long and ¼ inch wide at the middle, tapered at both ends (fig. 12-2). This wear leather will fit between the yoke and the ring, and it will be anchored in the center of the braid. Apply braiding soap to the strands and wear leather.

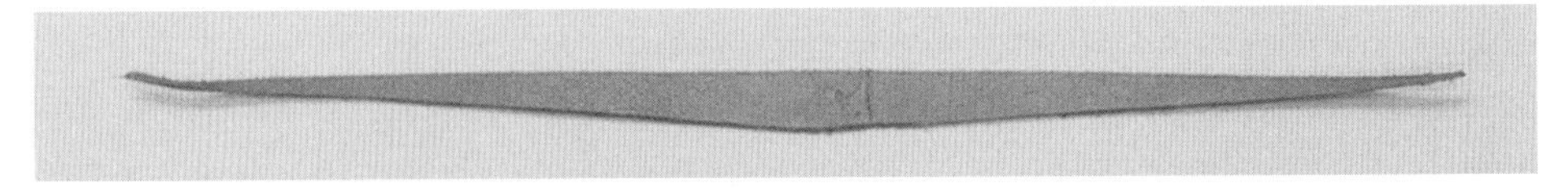

Fig. 12-2. Wear leather, 4 inches long and ¼ inch wide at the middle

3. Place the piece of wear leather over the key ring, and place the braided or solid yoke over the leather. Place the ring on the hook. This leaves one side of the yoke facing up and one side facing down. Arrange the three strands of the top side of the yoke so that two strands go to the left and one to the right. Arrange the three strands of the bottom side of the yoke so that two go to the left and one to the right. See figure 12-3.

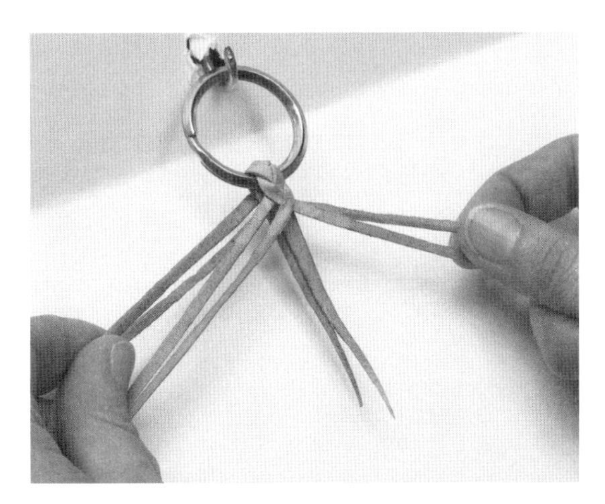

a. Wear leather and braided yoke is placed on ring.

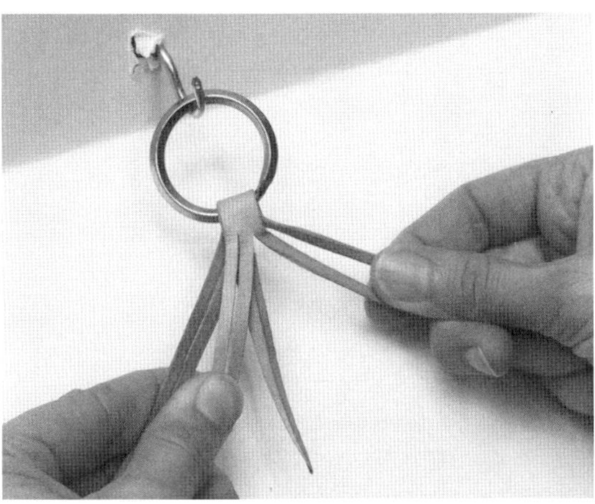

b. Wear leather and solid yoke is placed on ring.

Fig. 12-3. Wear leather positioned under yoke on rings

4. To set up the braid, the strands at the back are woven into those in the front, as shown in figure 12-4.

a. Take the upper left strand from the back and bring it around to the left to go under the upper of the two front left strands and over the lower.

b. Take the second strand at the left from the back and bring it around and over the upper of the two front left strands and under the lower.

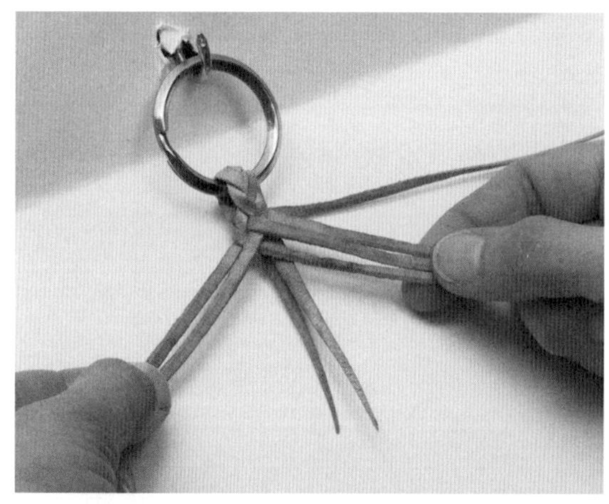

c. Take the right-hand strand from the back and bring it around to the right, under the upper strand, over the next, and under the last. This sets up the strands for a single-diamond (over-one, under-one) braid.

Fig. 12-4. Setting up a six-strand round braid

5. A single-diamond braid makes a neat start for the lanyard with either the braided or the solid yoke. It can be carried on for about 1½ inches. Follow the sequence in figure 12-5, working from the setup shown in figure 12-4c.

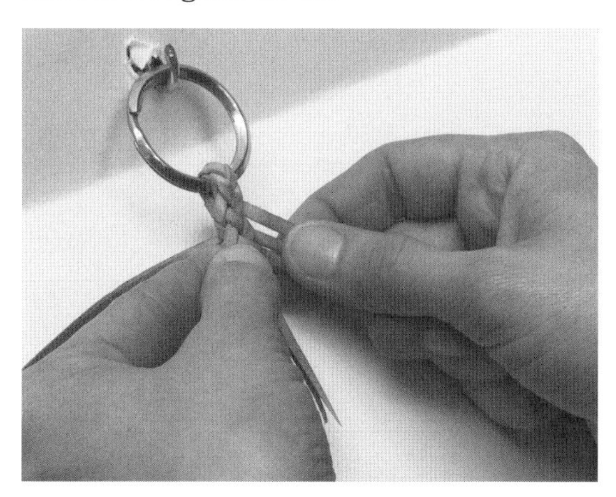

a. Hold with the left hand, one strand on the knuckle side and two on the palm side of the hand.

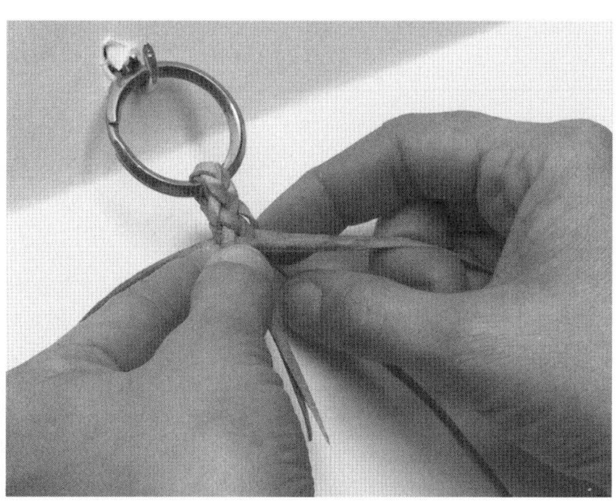

b. Depress the upper strand on the right, and lift up the next.

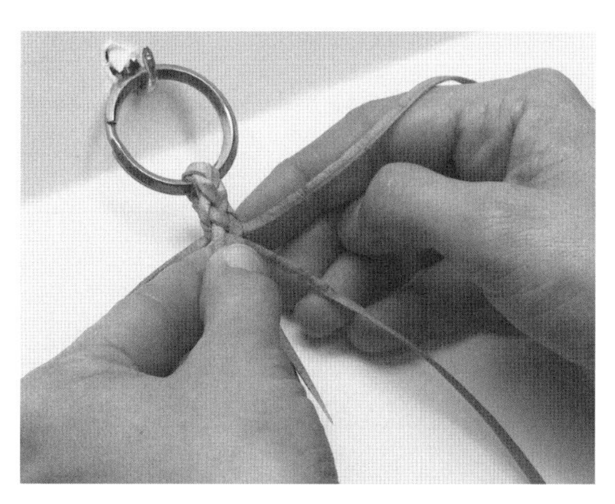

c. Reach through under the raised strand.

Fig. 12-5. Starting the lanyard with a single-diamond braid

Fig. 12-5. *Continued*

d. Bring the upper left strand through and pull it tight.

e. Lay this strand in place and change hands to hold with the right hand, the upper strand on the knuckle side and the other two on the palm side of the hand.

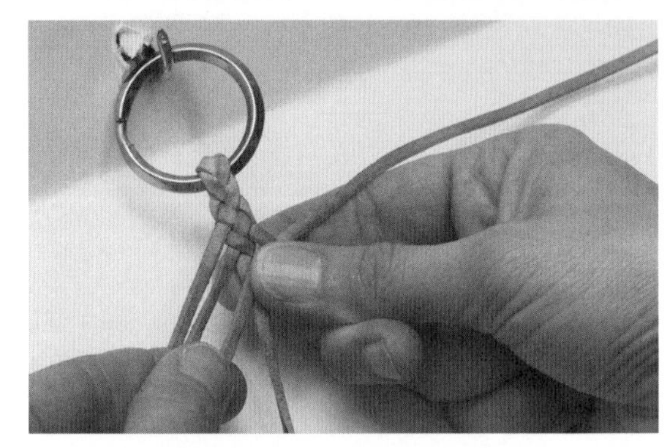

f. Depress the upper strand on the left.

g. Raise the next strand and reach through for the upper strand on the right.

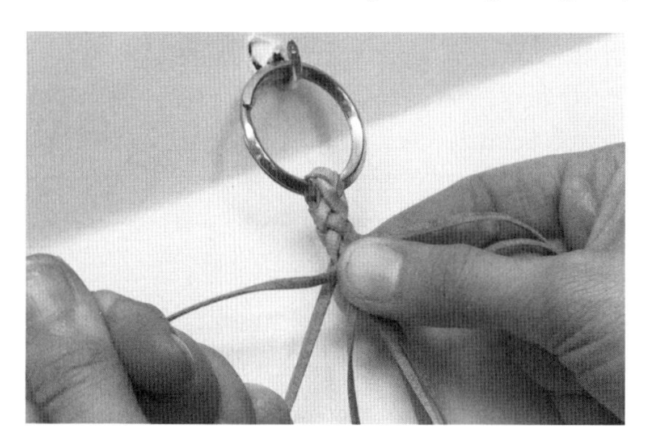

h. Bring this strand through, pull it tight, and lay it in place to complete the sequence.

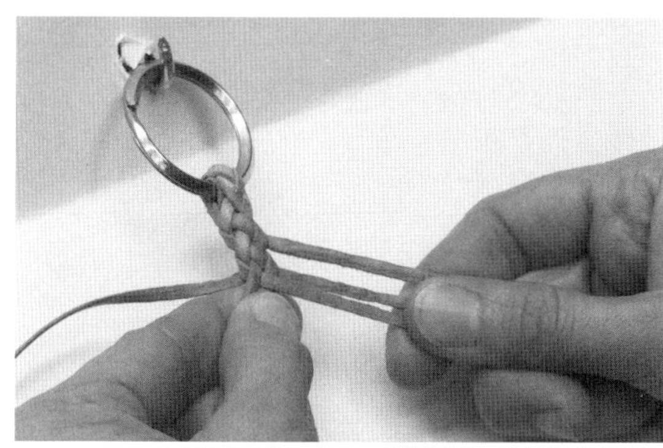

i. The sequence is complete. Diamond braid continues for about 1½ inches.

j. The single-diamond pattern ends, and four-seam braid can begin.

6. The main body of the braid will be done in regular four-seam braid, under-two and over-one with the right hand, under-one and over-two with the left hand. The transition from the single-diamond to four-seam is shown in figure 12-6.

a. Starting when the two center strands have been crossed right over left, hold with the right hand, all three strands on the knuckle side of the hand. Reach through with the left hand under the upper strand and over the two lower strands.

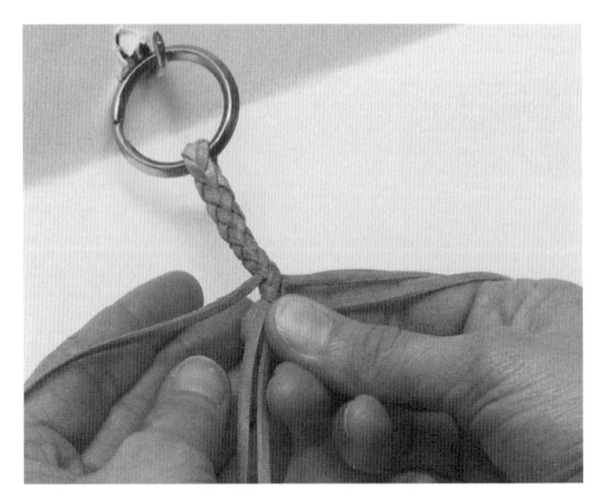

b. Bring the upper right strand through under the upper left and over the lower two left strands. Pull tight and lay in place.

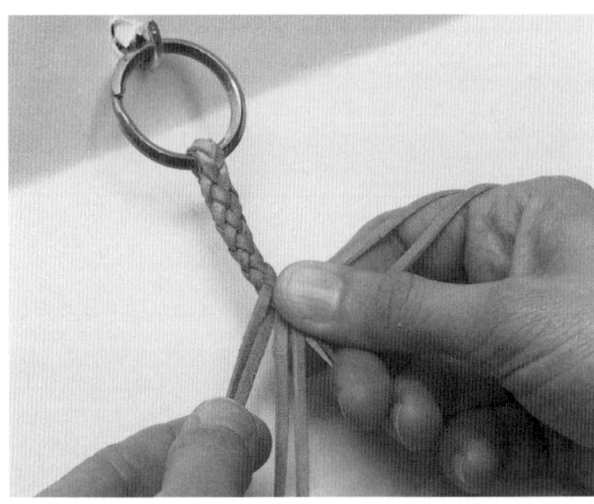

Fig. 12-6. Changing to four-seam pattern

c. Change to holding with the left hand, the two upper strands on the knuckle side and the lower strand on the palm side of the hand. The right hand is in place to reach through.

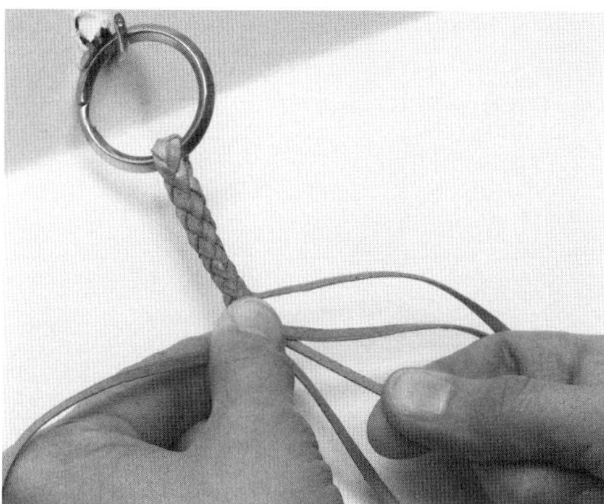

d. Bring the left upper strand through to the right, under the upper two strands and over the lower strand. Pull tight and lay in place.

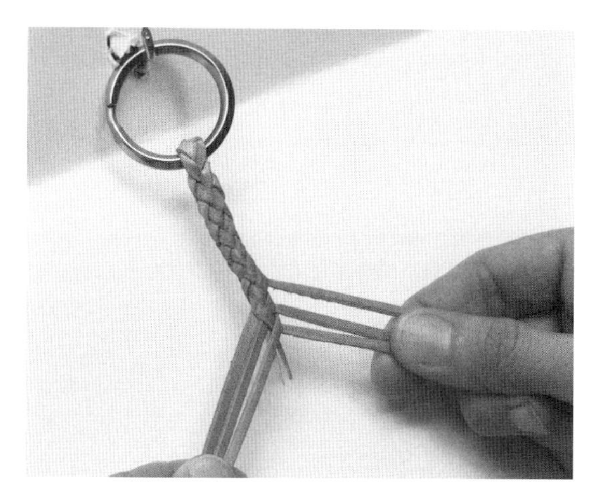

e. The change to four-seam is completed.

7. When the length from the ring is 15 inches, the braid will revert to single-diamond for 1½ inches. This single-diamond section will provide a good base for back-braiding the belt loop in place. The transition from four-seam to single-diamond is shown in figure 12-7.

a. Hold with right hand, the upper strand on the knuckle side, the two lower strands on the palm side of the hand.

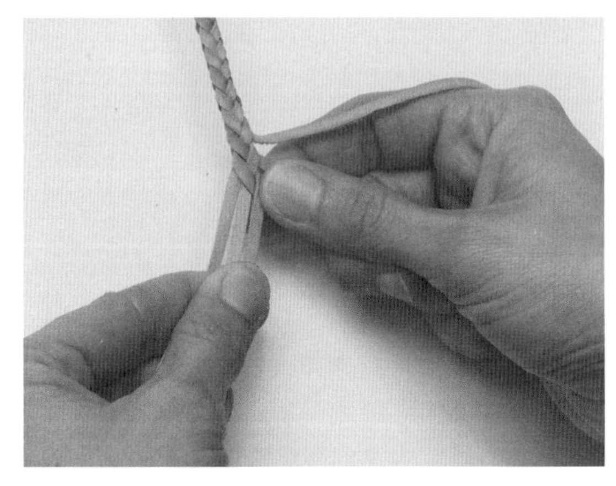

b. Depress the upper left strand, raise the next and reach through to bring the upper right-hand strand around.

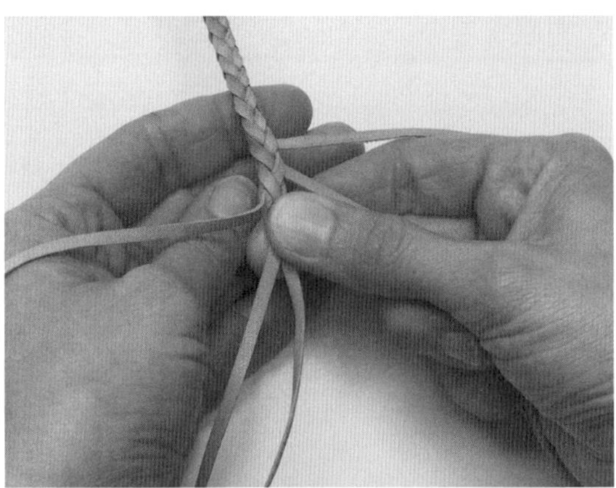

c. Bring the upper right-hand strand around and lay it in place as shown.

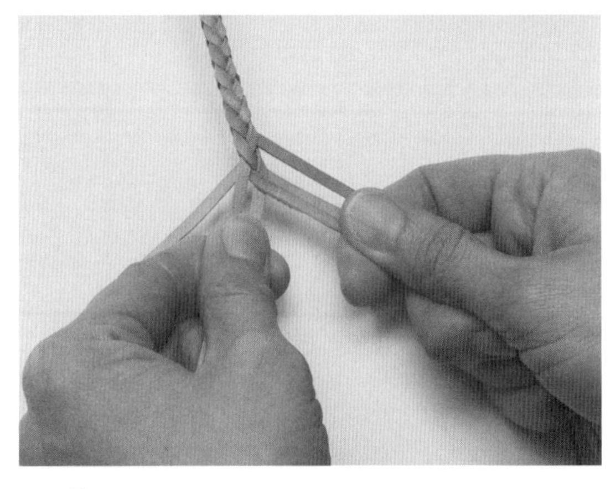

Fig. 12-7. Returning to single-diamond pattern

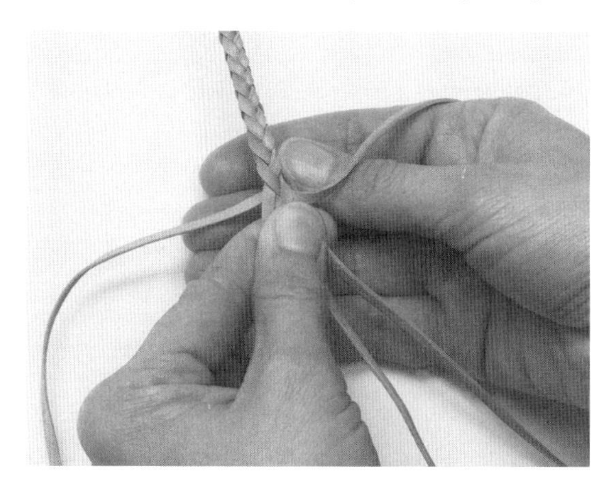

d. Hold with the left hand, the upper strand on the knuckle side, the two lower strands on the palm side of the hand. Depress the upper right strand, raise the next, and reach through with the right hand to bring the upper left strand through and into place to complete the change.

8. After the 1½-inch length of single-diamond pattern the braid is changed to flat braiding to make the loop. The transition is shown in figure 12-8. The flat braid is continued for 5 inches.

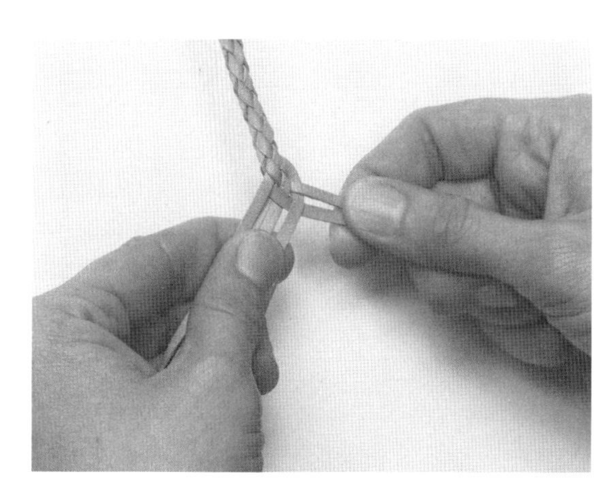

a. First tighten all strands, then bring the upper right strand under the next strand on the right and over the lower strand on the right.

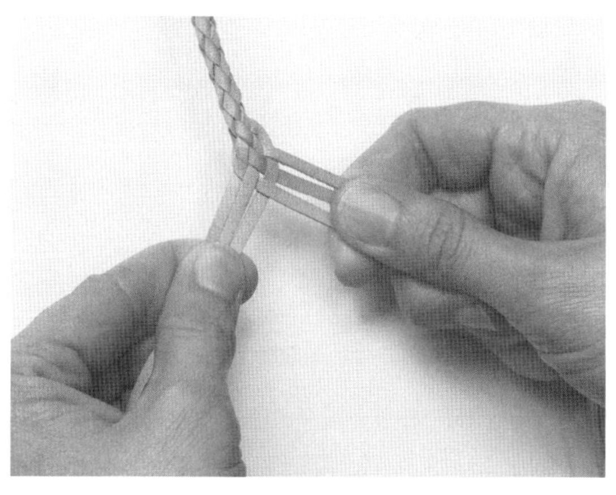

b. Bring the upper left strand over the next strand, under the next, and over the lower left strand, producing a regular flat-braid sequence.

Fig. 12-8. Changing to flat braiding

9. The flat section is then turned back to form the belt loop and back-braided into the round diamond braid. The back-braiding is shown in figure 12-9.

a. Point the ends of the strands and fold the flat braid back to make a loop. The fid shows the place where the first strand of the flat braid will be back-braided into the round braid.

b. Put the strand of the flat braid nearest to the loop under the indicated strand of the round braid, with the second strand over this strand and under the next strand in the round, and the third strand of the flat braid under the same strand of the round that the first went under, as shown.

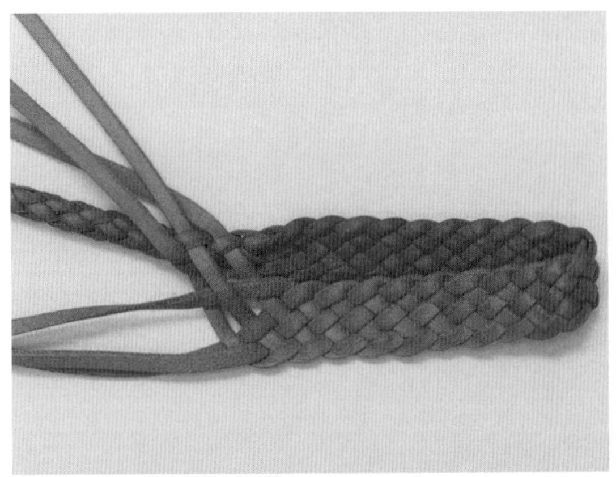

c. Tighten the first three strands and put the three strands on the left into the round braid. Start with the strand furthest from the loop, and put it under the strand immediately after it crosses the strand going to the right. Put the middle strand under the next strand of the round, and the last strand under the same strand as the first as shown. Tighten these strands.

Fig. 12-9. Forming the belt loop and back-braiding into the round diamond braid

d. Weave all strands through the round braid until the strands furthest from the loop have gone under three strands. Continue with the other strands to bring them up to the same as these, apart from the long strand which must finish just before the others.

10. The ends of the short strands are cut off fairly closely, and the long strand is used to make a covering knot over the ends of the strands cut off, as shown in figure 12-10.

The covering knot should be rolled lightly. If necessary, the ends of the knot may be squared up with a flat piece of wood or the side of a fid. Shellac the lanyard lightly, with a little extra shellac on the knot.

a. Cut off the short ends. The covering knot will be tied to cover the ends. Roll the lanyard, including the back-braided part.

b. Using the long strand tie a cover knot over the place where the ends are tied off.

Fig. 12-10. Cutting the strands and making a covering knot

FURTHER SUGGESTIONS

A six-strand key fob can be made with the same ring attachment, using a short length of braid and then a tassel. Similarly a watch fob can be made using a long yoke, narrow enough to attach to the bail of the watch with a lark's head knot. Lanyards, key fobs, and watch fobs can also be made in a four-strand braid, with lace cut to a central yoke.

CHAPTER THIRTEEN

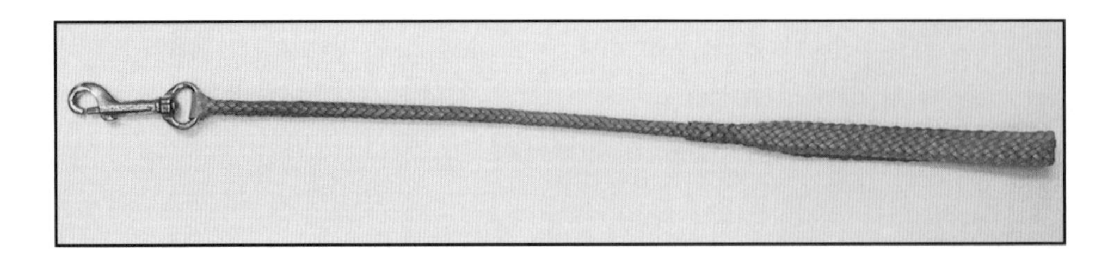

Project 4: Dog Lead, Eight-Strand Round Braid

This dog lead is made in an eight-strand braid over a leather core. Lace ⅛ inch wide may be cut from a medium-weight kangaroo skin for the overlay (leaving a central yoke), or ⅛-inch precut lace may be used with a braided yoke. The core may be of kangaroo or any soft thin leather. The bolt snap should be of good quality, with a hook opening of at least ¼ inch. A trigger snap may also be used.

1. Cut a core 27 inches long. The first 13½ inches of the core leather should be ⅜ of an inch wide. The last 13 inches should be ¼ inch wide, with a ½-inch taper between the two widths. In use the narrow half is rolled inside the wide half. To round the core, apply a good coating of braiding soap, tie the wide section to a horizontal ring, and round the narrow section flesh-side out by running a looped lace back and forth over it. Then put the core through the ring with the wide section, grain-side up, over the rounded narrow section, and round the entire core by pulling on a looped lace (fig. 13-1). The wide section of the core should pass just through the ring of the bolt snap to serve as a wear leather against the ring.

Precut lace may be used for the core if suitable thin leather is not available. Three ⅛-inch laces 27 inches long should be coated with braiding soap, middled, and placed on the ring of the bolt snap, two against the ring and one on top of these two. The resultant bundle of six strands should be rounded by running a looped lace back and forth over it in the usual fashion.

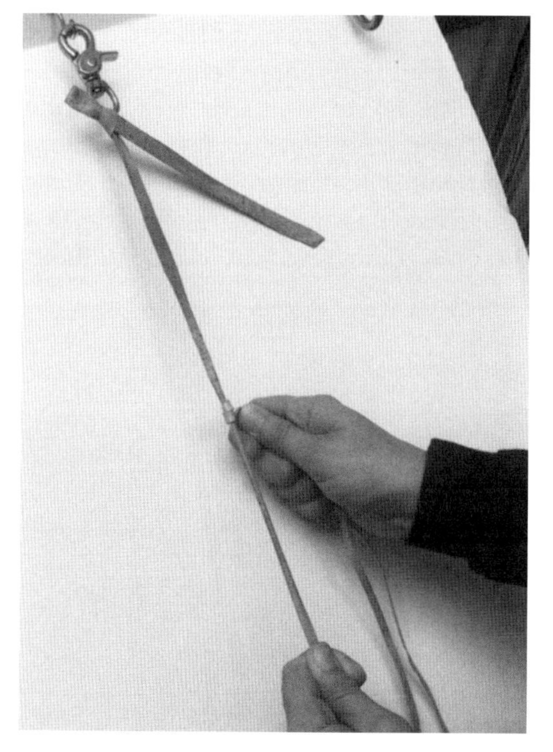

a. Round the narrow end of the core.

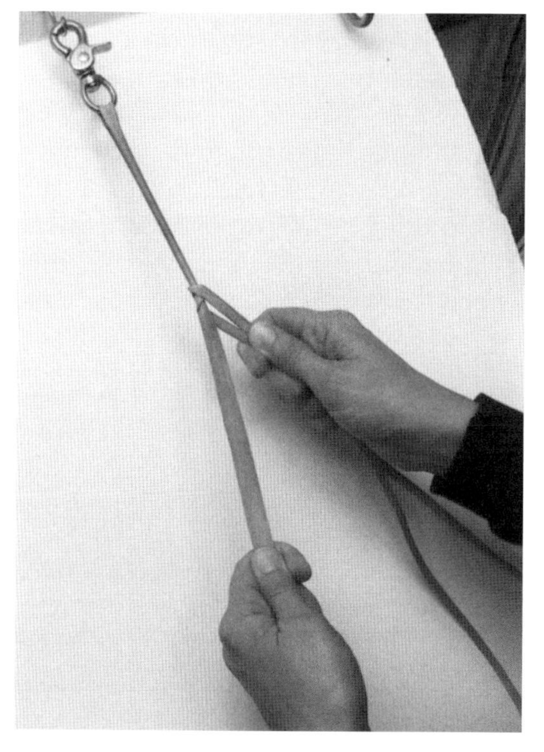

b. Round the entire core.

Fig. 13-1. Rounding the core for the dog lead

2. To cut a set of eight strands with a central yoke for the lead, start by cutting four strands 42 inches long, stopping them even with one another, and leaving them attached to the skin. Measure 7/8 of an inch beyond the 42-inch mark and start three more cuts so that four more strands can be cut leaving a 7/8-inch yoke. See figure 13-2. Cut these strands 42 inches long. Tie the set to a hook near the yoke one side at a time; pull the strands to test and stretch them, and pare the strands on opposite corners. Apply braiding soap.

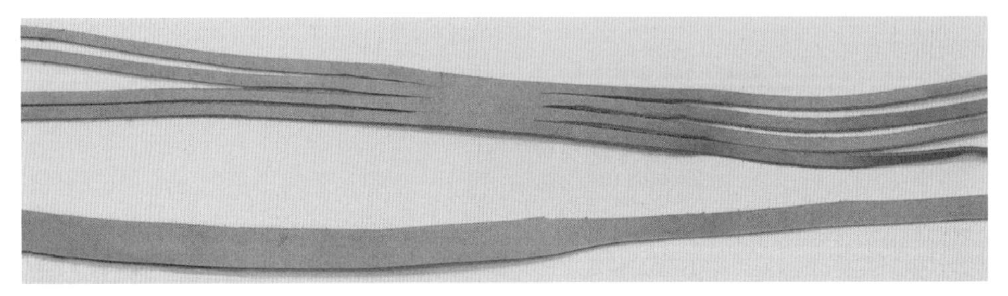

Fig. 13-2. Core and set showing solid yoke

3. To make a set from precut lace, use four strands 84 inches long. Tie them to the hook at the middle, pull to test and stretch them; pare the strands on opposite corners. Apply braiding soap and form a yoke at the middle about 7/8 of an inch long with a four-strand braid (fig. 13-3).

Put the core through the eye of the bolt snap, middled so the wide part of the core goes over the eye. Middle the set through the eye, with the yoke over the core. Hold the work so the ring is horizontal and the wide portion of the core is on top. One side of the yoke and its strands will be on top, and the other side will be underneath. The two half-sections of the core should form two concentric circles, each joining underneath, the smaller circle of the narrow portion filling the inside of the circle formed by the wide portion. The braid will start with the strands divided at the center, two out to each side, top and underneath.

Fig 13-3. Core and set showing braided yoke

4. A single-diamond braid is used at the beginning of this work. To start the braid from a solid yoke, follow the sequence in figure 13-4.

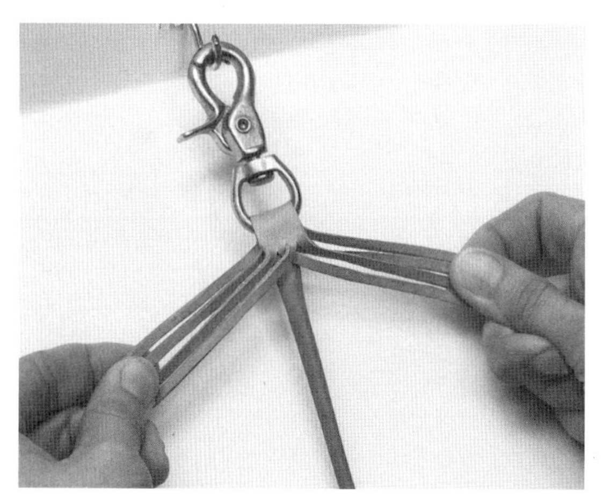

a. Cross the two center top strands right over left.

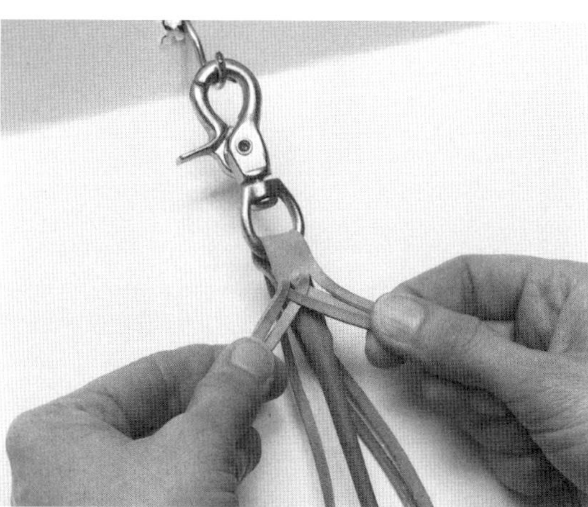

b. Take the upper left strand from the back of the yoke and bring it under the upper left top strand and over the next.

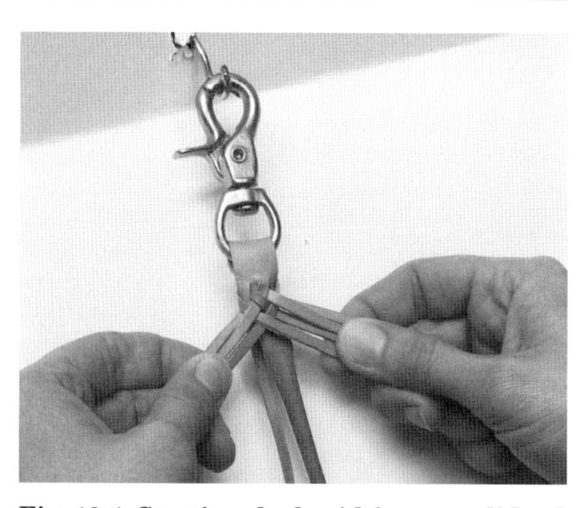

c. Take the upper right strand from the back of the yoke and bring it over the upper right top strand, under the next, and over the third strand. Now take the right-hand strand in the back of the yoke around from the back to the left, crossing under the left-hand strand remaining at the back of the yoke, over the upper left strand, under the next, and over the lower strand.

Fig. 13-4. Starting the braid from a solid yoke

Fig. 13-4. *Continued*

d. Take the remaining strand from underneath to the right and weave it under the upper strand, then over, under, and over the other three. The two center strands on the yoke underneath will now be crossed the same way and have the same appearance as the ones on the front, and the strands will be in the pattern of the single diamond.

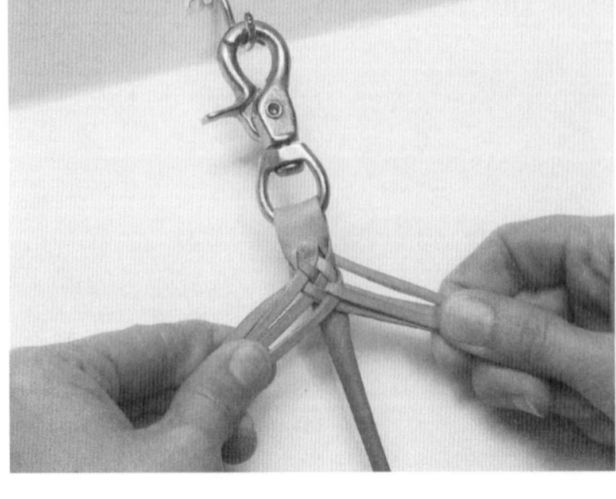

e. For the first stitch of the single-diamond braid, hold with the right hand, the upper strand on the knuckle side and the three lower strands on the palm side of the hand. Using the left hand, raise the upper left strand, depress the next, raise the next, and reach through to bring the upper right-hand strand around to the left and into place.

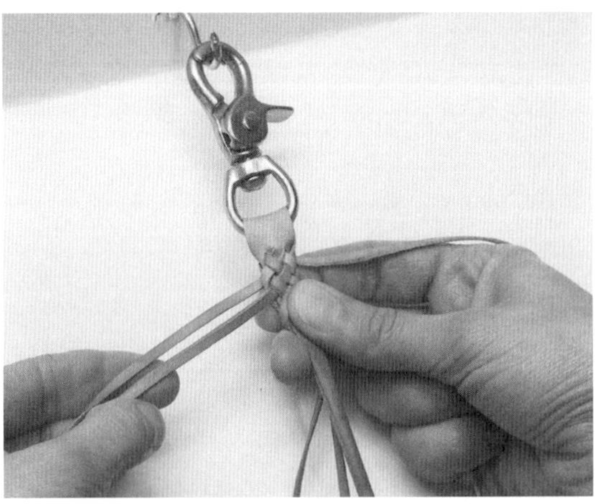

f. For the second stitch of the single-diamond braid, hold with the left hand, the upper strand on the knuckle side, the lower strands on the palm side of the hand. Using the right hand, raise the upper right strand, depress the next, raise the next, and reach through to bring the upper left strand around to the front and into place. This completes a full sequence. Continue with the single-diamond work for about 2 inches.

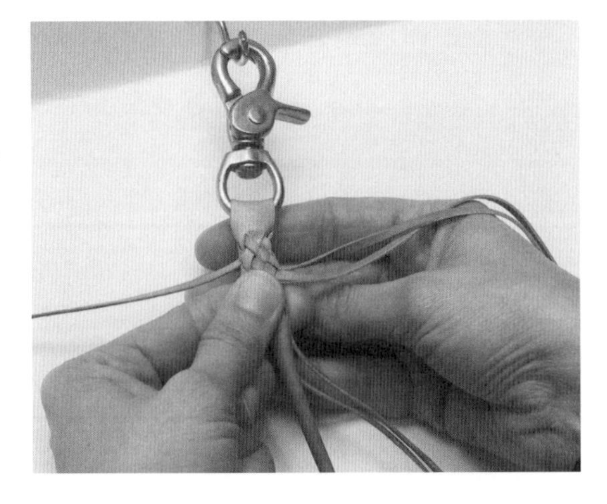

5. To start the braid from a braided yoke follow the steps shown in figure 13-5.

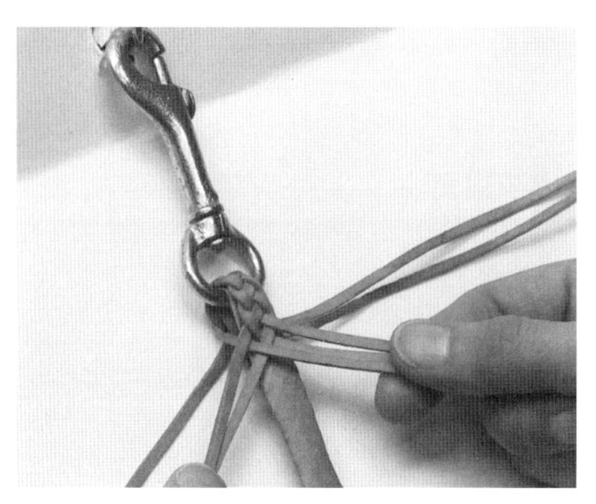

a. Bring the upper left strand from the back of the yoke under the upper left strand and over the lower strand on the front.

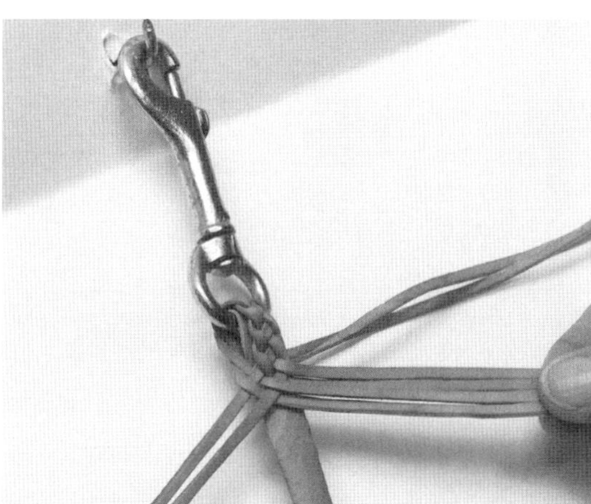

b. Bring the remaining left strand from the back of the yoke over the upper and under the lower strands on the front left.

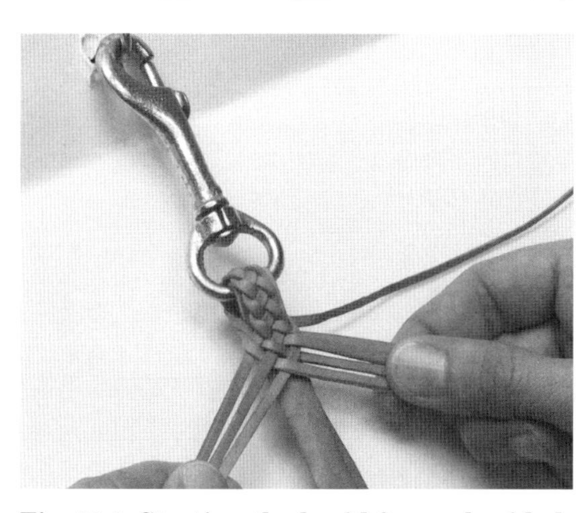

c. Bring the upper right strand from the back of the yoke over the upper right strand on the front, then under, over, and under the next strands.

Fig. 13-5. Starting the braid from a braided yoke

Fig. 13-5. *Continued*

d. Bring the remaining strand from the back of the yoke under the upper right strand on the front, then over, under, and over the other strands. The pattern for single-diamond work is now set up.

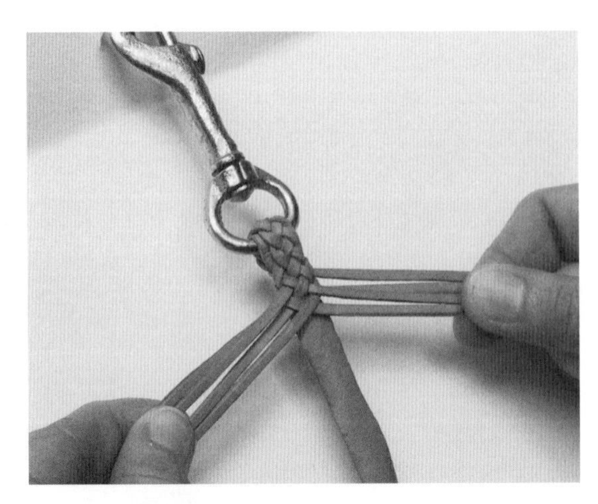

e. For the first stitch of the single-diamond braid, hold with the right hand with one strand on the knuckle side of the hand. Using the left hand, raise the upper left strand, depress the next, raise the next, and reach through to bring the upper right around to the left and into place.

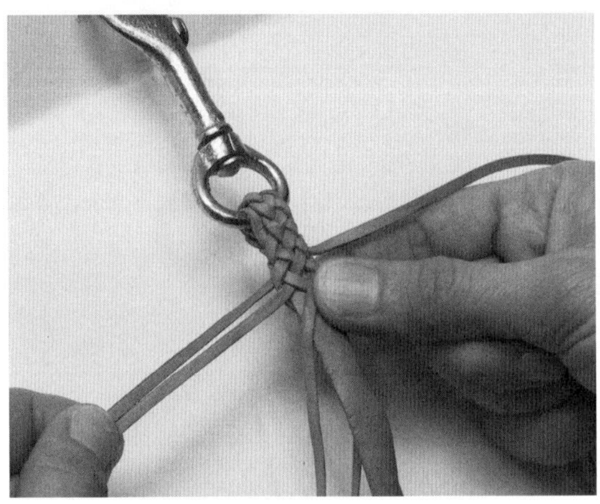

f. For the second stitch of the single-diamond braid, hold with the left hand, with the upper strand on the knuckle side of the hand. Using the right hand, raise the upper right strand, depress the next, raise the next, and reach through to bring the upper left strand around to the right and into place. This completes a full sequence. Continue with single-diamond work for 2 inches.

6. At about 2 inches change to a regular four-seam braid—under-two, over-two—as shown in figure 13-6.

a. To change to four-seam, start with the center strands crossing right over left. Hold with the right hand, the three upper strands on the knuckle side, the lower strand on the palm side of the hand. Reach through with the left hand under the upper two strands on the left and over the lower two to bring the upper strand from the right around under the two upper and over the two lower strands on the left and into place.

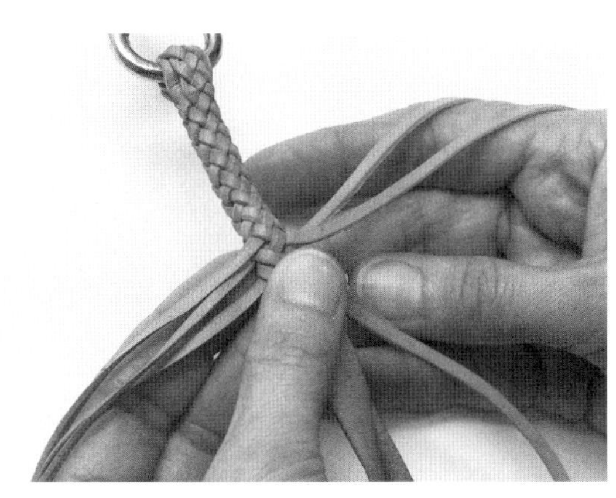

b. Holding with the left hand, three strands on the knuckle side, reach through with the right hand and bring the upper strand from the left around and into place.

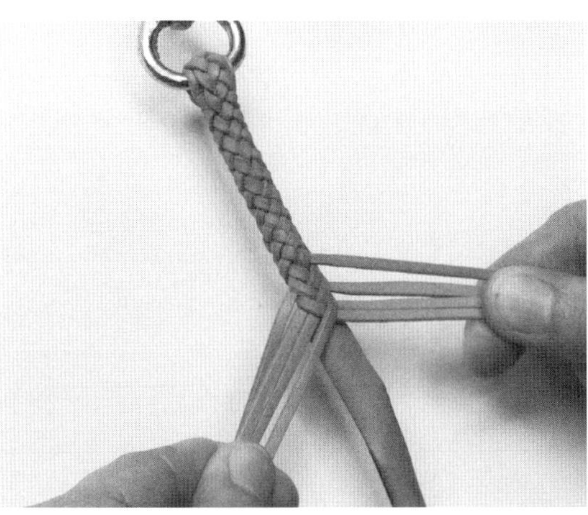

c. Complete the change to four-seam. Repeat steps one and two, taking care to pick up the correct strand. Since the strands are coming out of the diamond pattern, the correct strand is not so obvious as in the regular four-seam. After this repetition the change is complete, as shown. Continue with the four-seam work to 9 inches of braid.

Fig. 13-6. Changing to regular four-seam work

7. At 9 inches revert to single-diamond work for 3 inches. This section of single-diamond will serve as the base for back-braiding the wrist loop into place. See figure 13-7.

a. To change to single-diamond, start with the center strands crossing left over right. Hold with the left hand, the upper strand on the knuckle side. With the right hand raise the upper right strand, depress the next, and raise the next. Reach through under the two raised strands, and bring the upper left strand through and into place.

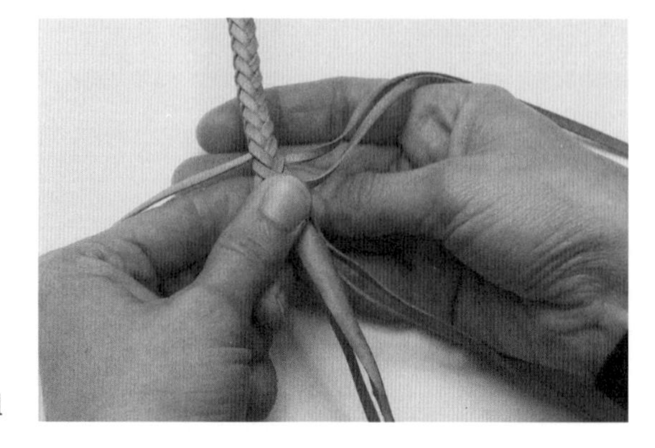

b. Hold with the right hand. With the left hand, raise the upper left strand, depress the next, and raise the next. Reach through under the two raised strands and bring the upper right strand through and into place. Continue the single-diamond work for 3 inches.

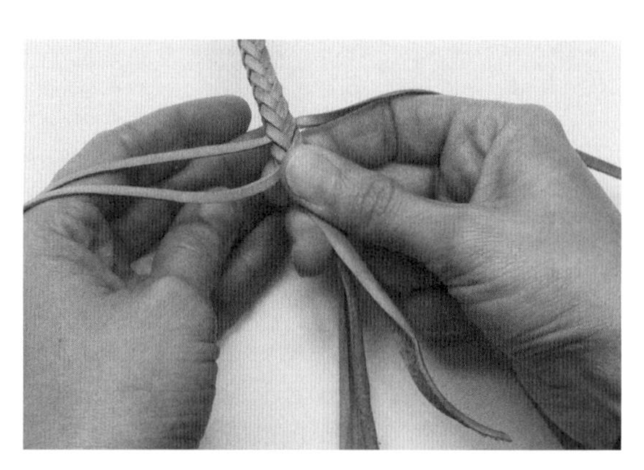

Fig. 13-7. Reverting to single-diamond work

8. After 3 inches of diamond braid, change to a flat braid for the wrist loop. See figure 13-8. Carry on the flat braid for 13½ inches.

a. To change to flat braid, first tighten all strands in the round braid. With the center strands crossed right over left, take the upper left strand and weave it over, under, and over the other left-hand strands.

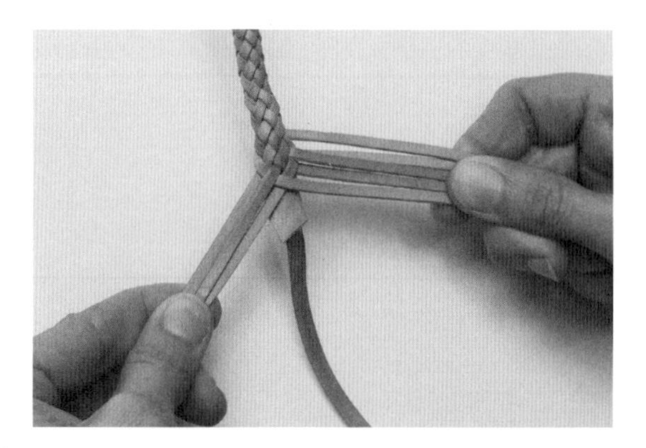

Fig. 13-8. Changing to flat braid

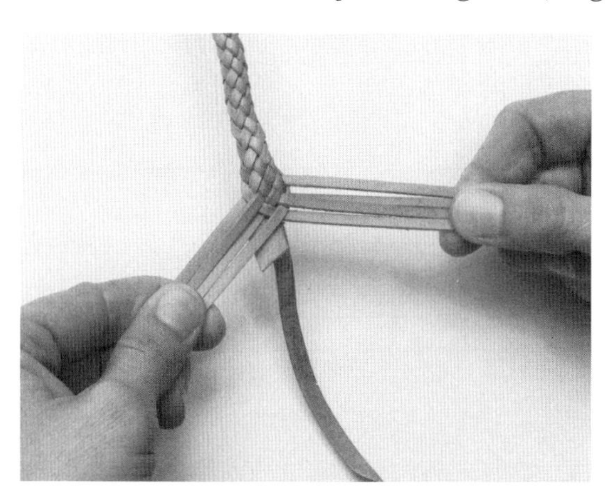

b. Take the upper right strand and weave it under, over, under, and over the other right-hand strands. Continue this flat braid for 13½ inches.

9. To form the wrist loop, turn the flat braid back and back-braid the end into the single-diamond section at the end of the round braid. See figure 13-9.

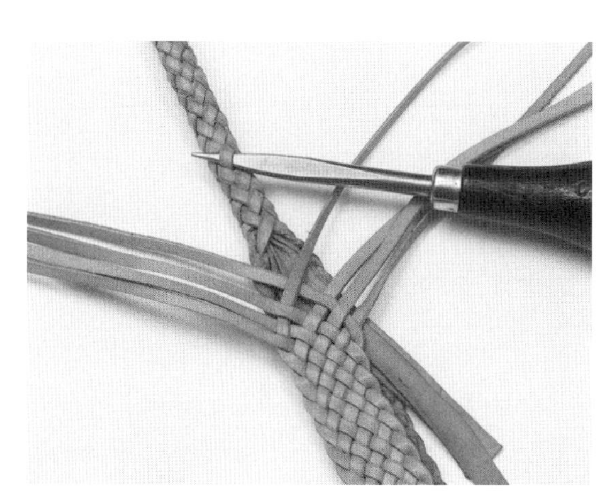

a. To back-braid the wrist loop, fold the flat braid, grain-side out, back to the round braid. Select the fourth crossing of the center strands from the end of the round braid as shown by the position of the fid.

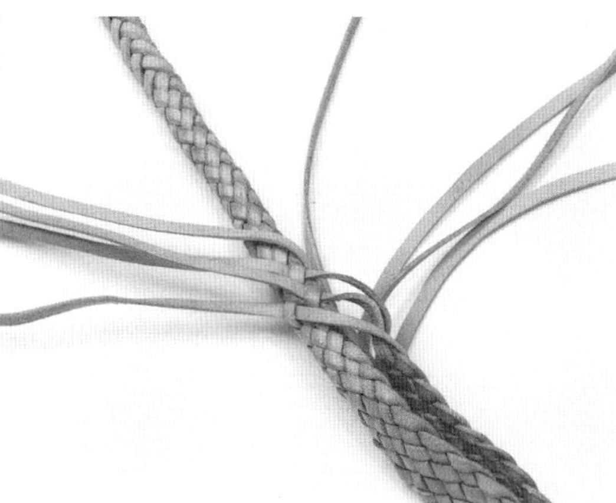

b. Insert the upper left strand of the flat braid under the top strand of the selected cross as shown. Then insert the second left strand of the flat under the adjacent strand, the third strand of the flat under the same strand of the round braid as the first, and the fourth under the same strand of the round braid as the second.

Fig. 13-9. Forming the wrist loop

Fig. 13-9. *Continued*

c. Take the upper right strand of the flat braid over the selected cross and insert it under the adjacent strand of the round braid. Insert the second right strand of the flat under the strand leading into the selected cross, the third under the same strand as the first, and the fourth under the same strand as the second. Tighten all eight strands uniformly.

d. Continue inserting the free ends so that each traces the strand on top of which it now lies. Carry each strand under at least four strands and bring the braid to a symmetrical finish. Trim the ends to about ⅛ inch, and trim off the excess core inside the loop.

To finish the lead, roll the round braid. Take care to have the snap over the edge of the rolling board, free to move when rolling. Roll the back-braid, but not the flat. Then shellac the round and grain side of the loop.

FURTHER SUGGESTIONS

A more ornate dog lead can be made by putting a covering knot over the back-braided ends. Dog leads can be made in six-strand braid or twelve-strand braid. Open reins or horse leads can be made in a similar fashion.

CHAPTER FOURTEEN

Project 5: Hatband, Four-Strand Flat Braid

Flat braids make attractive hatbands. This four-strand flat braid band is adjustable to a wide range of hat sizes. It may be made from precut lace or cut to a yoke to form an attractive fringe at the end underneath the band. With precut lace, use two ⅛-inch strands 84 inches long. For a band started from a yoke, cut four ⅛-inch strands 42 inches long to a yoke of about 4 inches long. The lace for hatbands is not pulled heavily to test it or to take the stretch out. Flat bands are braided with moderate tension, so that when put around a hat they can adjust better to the hat.

1. Tie the yoke to the hook or middle the precut strands and braid 31 inches in flat braid (fig. 14-1). Dust with talcum powder as a lubricant. (With hatbands, talcum powder is best; avoid grease that might stain.)

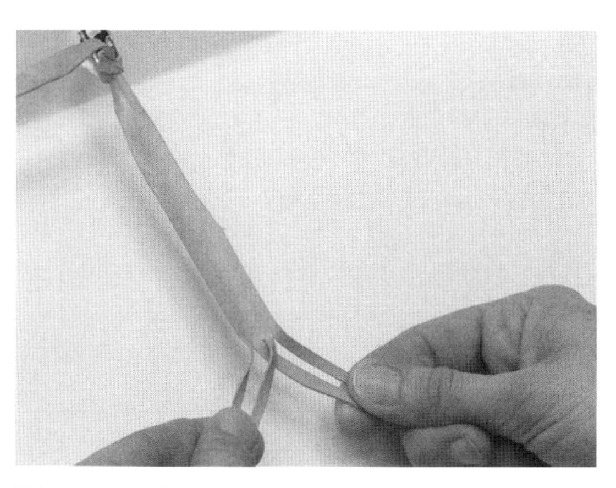

a. Tie yoke to hook and cross center strands, right over left.

Fig. 14-1. Flat braiding a hatband

Fig. 14-1. *Continued*

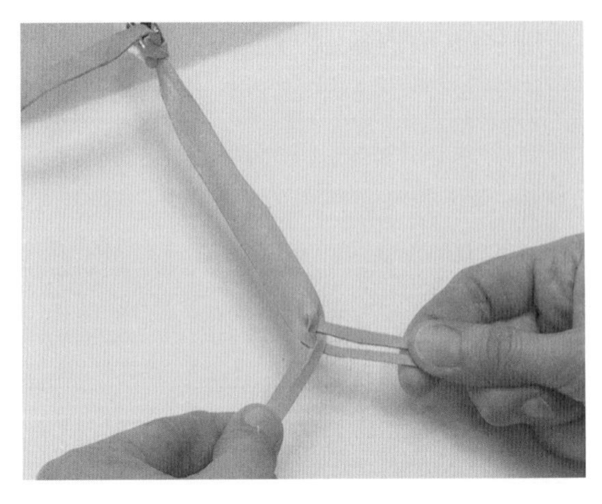

b. Weave in the pair of outer strands. This gives the start of a regular four-strand flat braid.

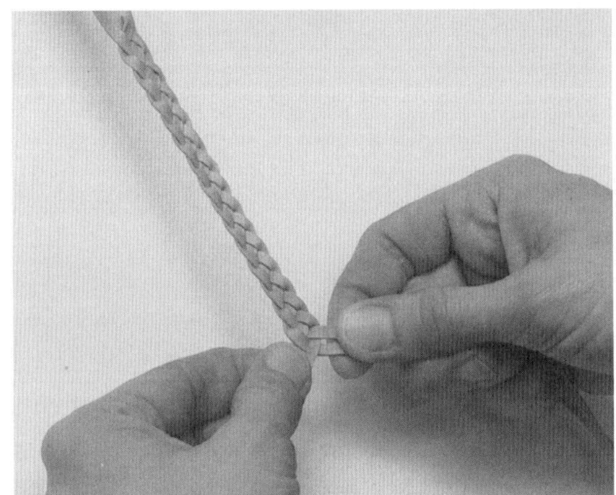

c. Continue braiding to 31 inches. The tension should be only moderately firm, but even throughout.

2. At 31 inches tie off the end with two half hitches. See figure 14-2.

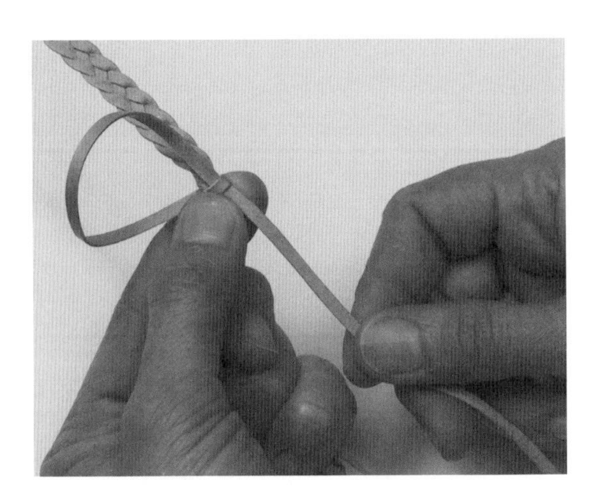

a. Make a half hitch. Take the upper right-hand strand around to the back and then over the front. Put the end through at the back, tighten by pulling the loop firmly across the front, and then bring the end into place.

Fig. 14-2. Tying off the end

b. Make a second hitch beside this hitch on the fringe side and over the end of the first hitch using what was the upper left strand. Again, tighten the loop over the end of the strand before the end is pulled into place.

c. The two hitches should be close together and neat. Trim the ends to about one inch.

3. If the band has a yoke, the yoke can be cut to about 1½ inches and the end split into four strands as shown in figure 14-3.

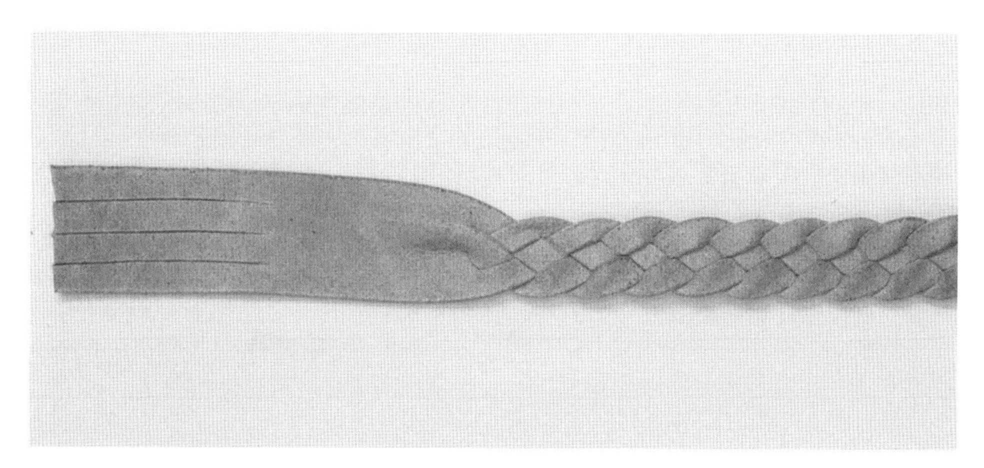

Fig. 14-3. Forming fringe on the yoke

4. To assemble the band, turn it into a circle with a few inches of overlap. Using ⅛-inch lace, put on two knots of the same type used to cover the tassel end of the whistle lanyard. These knots will hold the band and allow adjustment to size. They should be made to fit snugly, but without crushing the braid. See figure 14-4.

Shellac lightly on the grain side of the band. Shellac the knots more heavily, including the underside.

a. Start the knots on the inside of the band.

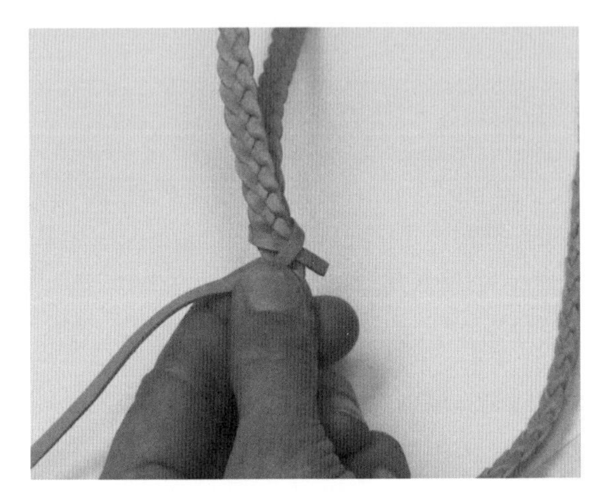

b. Finish the knots by trimming the ends of the lace so that they will not show from the front of the band.

Fig. 14-4. Assembling the hatband

CHAPTER FIFTEEN

Project 6: Hatband, Ten-Strand Flat Braid

A wider ten-strand flat braid hatband is attractive on many hats. This band has two single-strand keepers—strands going around the ends—to keep the two ends of the band together while allowing them to slide one against the other. The band can be adjusted to fit a wide range of hat sizes. It requires 5 pieces of lace, each 84 inches long.

1. Middle all strands on the hook and stretch them lightly. Sprinkle them with talcum powder for lubrication. Start the braid on the hook in the usual fashion as shown in figure 15-1. The first 3 inches should be braided tightly enough to help hold the keeper at that end, but the tightness of the remainder of the braid should be only moderate, so that the band can adapt better to the hat shape.

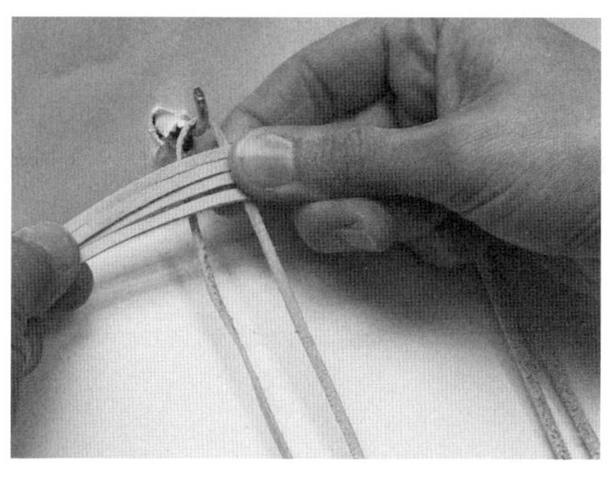

a. Hold four middled strands over a strand middled on the hook.

Fig. 15-1. Starting the ten-strand flat braid

Fig. 15-1. *Continued*

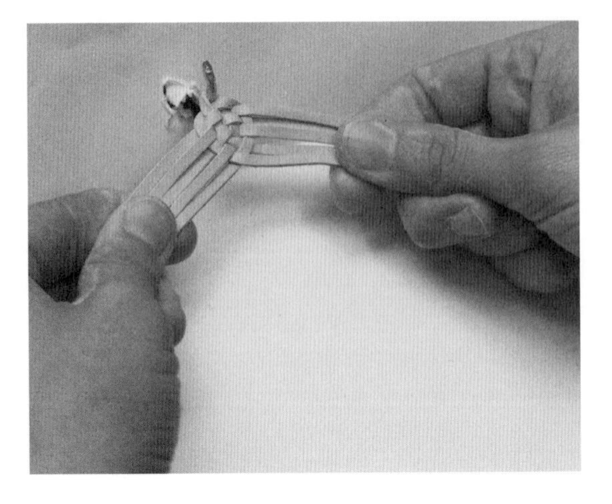

b. Weave upper strands through and tighten the braid up close to the hook.

2. Braid to a length of 31 inches, hold the end in a round shape, and tie off with two half hitches (fig. 15-2). Use the upper right-hand strand for the first hitch, and put the end on the underside to keep it neat. Use the upper left-hand strand for the second hitch. Trim the ends to random lengths of two to three inches.

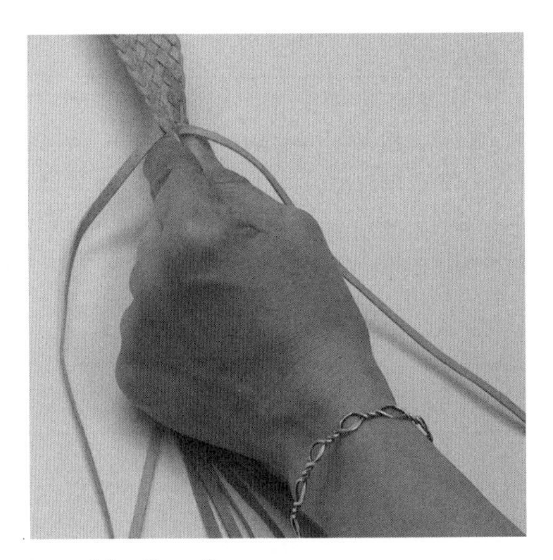

a. With the band flesh-side up, hold the fringe with the braid rounded and the top strand on each side free. The strand for the first hitch will now be on the left.

Fig. 15-2. Tying the fringe on the ten-strand hatband

b. Take the top left-hand strand and loop it around the braid and under itself, with the end placed flesh-side up, making a half hitch around the braid.

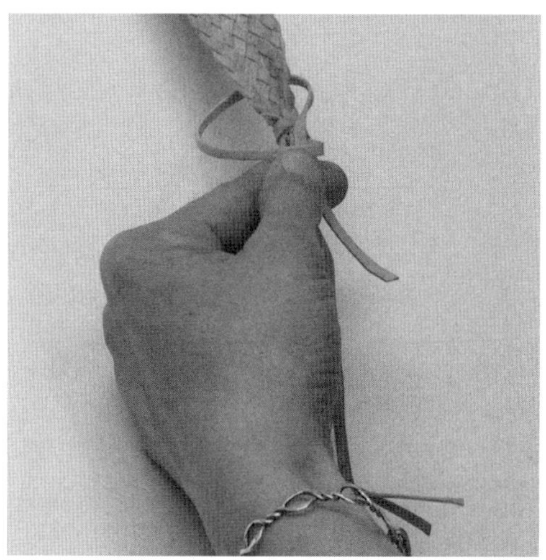

c. Tighten this hitch by pulling the loop tightly over the end before pulling on the end. Then make a similar loop with the top right-hand strand.

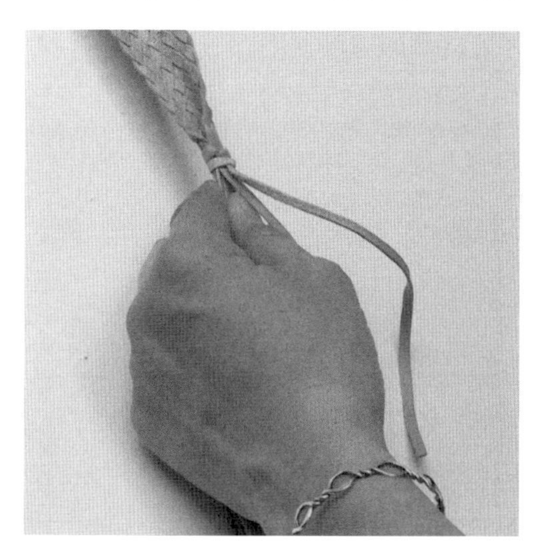

d. Tighten this hitch by pulling on the loop before pulling the end into place, which completes the tie.

3. Cut two strands of lace about 12 inches long for keepers, and thread a needle onto one of them. Find a strand in the braid close to the tassel end but not where the end is rounded. Follow this strand back two widths of the braid, and insert the needle to begin the first keeper. Work the end of the lace into the flesh side at this point, and weave the keeper strand first across the band to the opposite side, and then back across the band again, following the selected strand of the braiding (fig. 15-3). Take the keeper lace around to the other side of the braid to form the actual keeper. To secure the keeper, work the lace from this side across the braid, leaving the keeper loose. Leaving the lace loose at the side, again weave it across the braid back to the side opposite the starting point.

Fig. 15-3. Setting up the first keeper

4. Insert the starting end of the braid about 6 inches and tighten the keeper loop by pulling on the loose loop at the side and then pulling the end of the lace into place. Make the keeper snug but not so tight as to distort the band. See figure 15-4, which shows the grain-side (outer side) at this section.

Fig. 15-4. Finishing the first keeper

5. Make a similar keeper at the starting end of the band, working the lace into the flesh-side of the braid and running the keeper around the outside to hold the two parts of the band together. See figure 15-5.

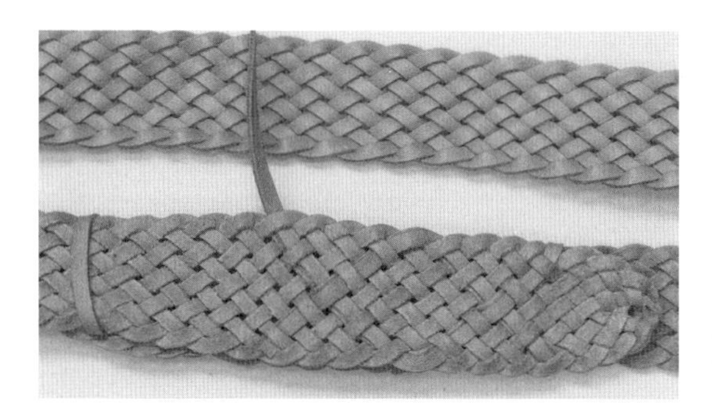

Fig. 15-5. Inserting second keeper

Turn the band so the grain-side is out, and shellac lightly, with a little extra shellac on the flesh-side where the keeper lace is worked in.

FURTHER SUGGESTIONS

Eight- or twelve-strand hatbands can be made in a similar fashion. Bands can be made using two-tone or fancy braids.

CHAPTER SIXTEEN

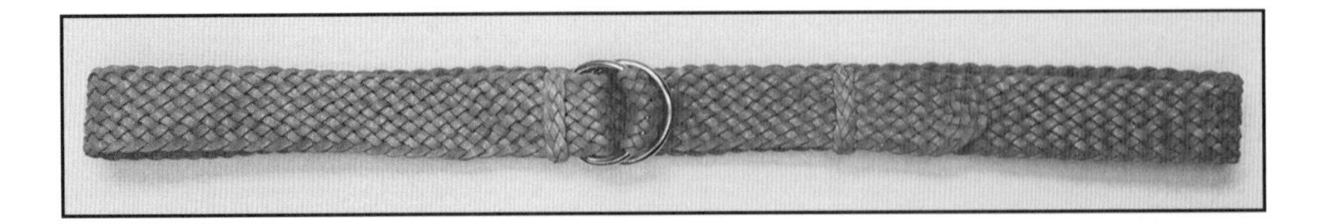

Project 7: Belt, Twelve-Strand Flat Braid

The traditional Australian braided belt has a cinch ring fastening made from two harness rings or dees. The belt illustrated is made in twelve strands of ⅛-inch lace, producing an overall width of about 1 inch. The dees are l inch across the flat. Only the best, firmest lace should be used in belts, and it should be braided firmly. A belt to fit a 34-inch waist will need 6 pieces of lace, each 112 inches long. For every 2 additional inches in waist size, add 6 inches to the lace length.

1. Tie the strands to a hook and pull them strongly to stretch them fully. If the strands are not well stretched the belt may later stretch badly and become uneven. If the strands stretch and become noticeably uneven, they should be used for some other project, and firmer, more even strands selected for the belt. Apply a coating of braiding soap.

Middle the strands, put one strand on the hook, and add on the rest in the usual fashion. In pulling the strands for the first few passes, braid firmly and tightly but do not pull so tightly as to distort the pointed end of the braid (fig. 16-1). Throughout the braiding, pull the strands firmly before working them, but not so much as to distort the edge on the opposite side. After working a strand, again pull it firmly into place. In both cases, pull the strand in the direction that it is running. The objective is to make a very firm braid, but also very even.

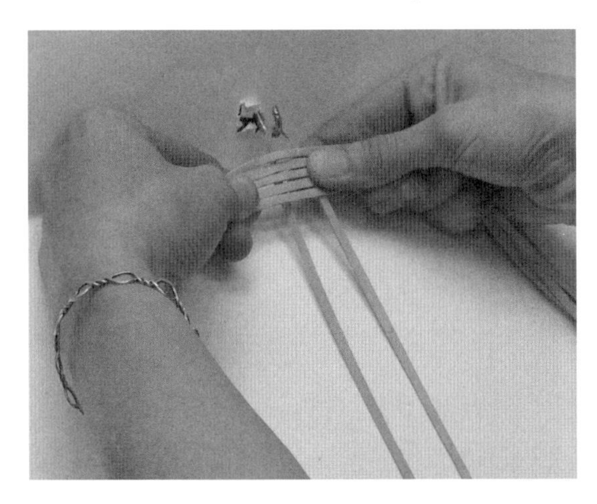

a. Middle the six strands; place one middled strand on the hook and hold the other five over this strand.

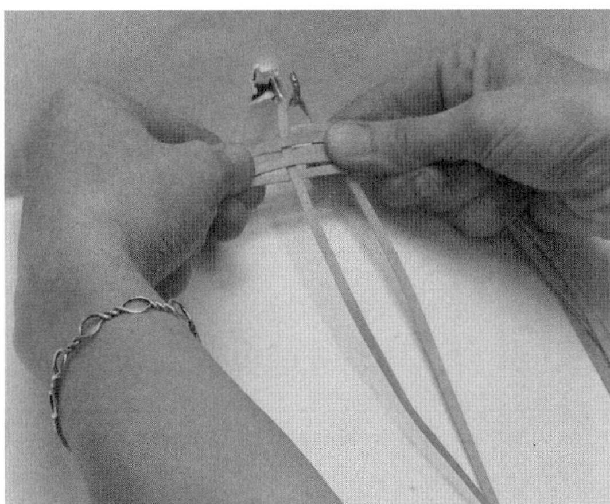

b. Weave the left-hand part of the strand on the hook over, under, over, under, and over the other five strands.

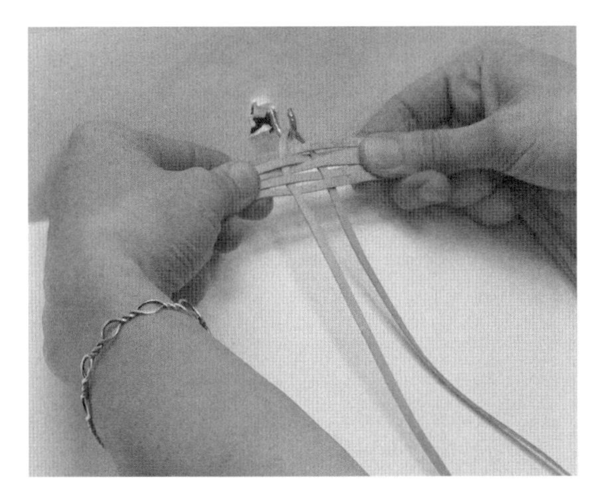

c. Weave the right-hand part of the strand on the hook under, over, under, over, and under the five strands.

Fig. 16-1. Starting the twelve-strand flat braid

Fig. 16-1. *Continued*

d. Cross the strands coming from the hook, right over left, and bring the assembly neatly together.

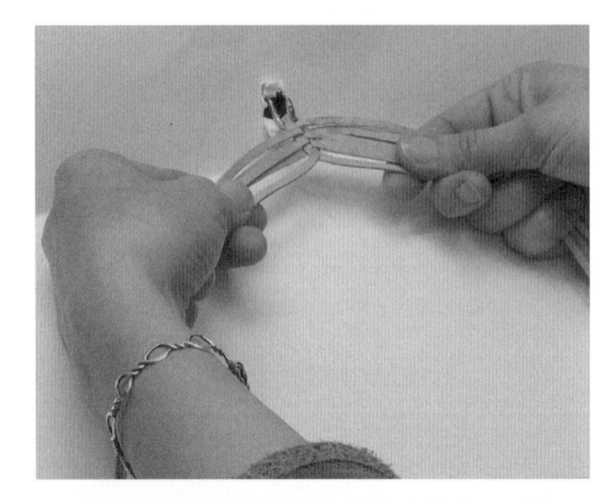

e. Weave the first of the five strands through each side, on the right going under the next strand, on the left going over the next strand.

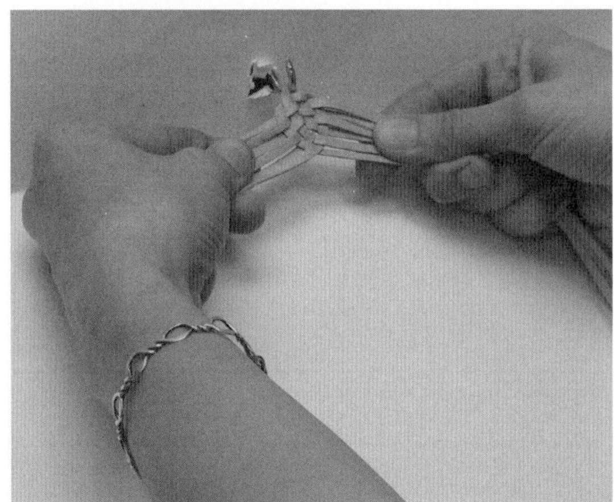

f. Repeat this procedure through the length of the braid. Keep the braid tight by pulling each strand both before and after it is woven through.

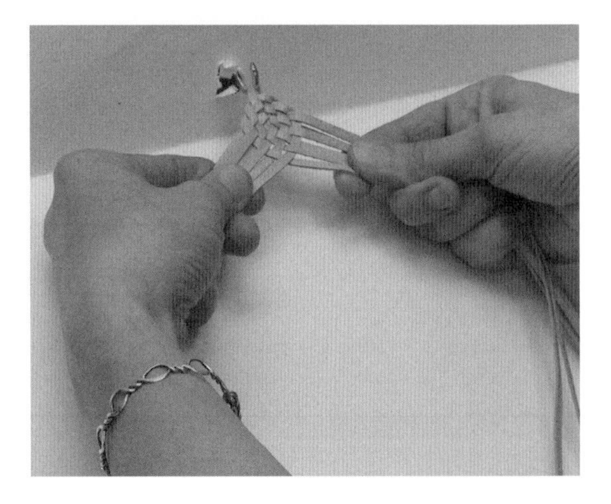

2. After braiding 9 inches, turn over each strand just before it is worked through, until all the strands are turned. This will reverse the grain-side of the leather on the end of the belt, so that only the grain-side will show when the end is folded back through the rings. Continue braiding for the full length. See figure 16-2.

a. Turn the first strand at the edge, and weave it through flesh-side up.

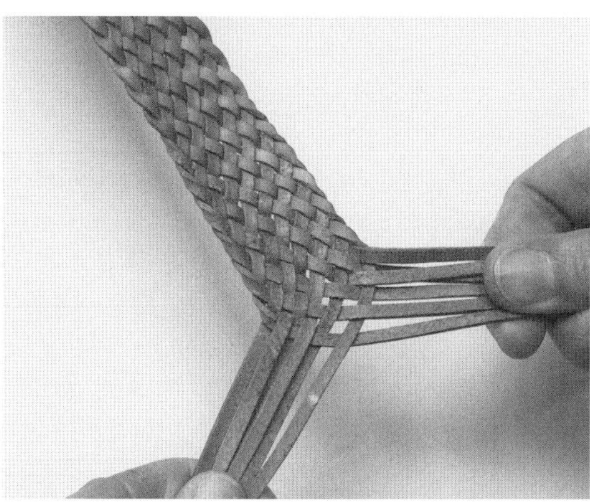

b. All strands are turned and braiding can be continued in a normal manner. The belt, of course, will now be flesh-side up as it is being braided.

Fig. 16-2. Turning the strands

3. During braiding, the first strand at the beginning end of the belt may stretch into a loop where it is held on the hook. If this occurs, tighten up the loop at the end by pulling on the strands where they cross, then bring the strands into place at the edges if they have been pulled in. See figure 16-3.

a. Pull out the loop from inside the belt.

b. Adjust the strands at the side if they have been pulled out of line.

Fig. 16-3. Removing a loop from the end of the belt

Before measuring for length, take out any stretch in the belt by holding both ends in the hands and pulling the belt against your foot or around a post. Do not pull from the hook to stretch the belt, as this could break the strand holding the belt to the hook.

4. To attach the dees, first point the ends of the strands, and reeve the belt through the rings. To set the desired waist size, turn the start of the belt back so that the grain-side on the start just covers the flesh-side where the strands were turned. Adjust the end through the rings so that the length measures the waist size, excluding the rings. Unbraid the loose end of the belt to within an inch or so of the rings, and back-braid the ends in the sequence illustrated in figure 16-4. When the ends are back-braided, do not pull them up too snugly lest they distort the front of the belt; pull them only just into place. To finish, trim off the ends to about ⅛ inch.

a. With the end of the belt passed through the dees and folded over the body of the belt, insert the lower right strand of the belt end under a strand on the edge of the body about ½ inch from the dees. Leave the next strand on the right free, but insert the remaining strands on the right under succeeding strands second from the edge as shown. Pull all strands into place so that the strands on the end of the braid are all tracking appropriate strands in the body of the belt.

b. Starting with the upper left strand, insert the first five strands under the last strands on the left edge. Leave the lower strand free. Make sure that the upper strand of the end of the belt is tracking the same strand as the body of the belt from the right edge, and note that the fifth strand will go under the last strand of the end as well as the strand on the left edge of the body.

c. Take the lower end on the right, put it over the next end, and insert it under the first strand of the body of the belt, over one double strand (i.e., over both the strand of the end and the tracking strand in the body of the belt), under just the strand from the belt end but not under the strand in the body of the belt, over the next strand from the belt end and under the next strand in the body of the belt.

Fig. 16-4. Attaching the dees

Fig. 16-4. *Continued*

d. Take the next, now lower, end on the right, insert it under the first strand in the body of the belt, over the next double strand, under just the strand from the belt end but not under the strand in the body of the belt, over the next strand, and under the next.

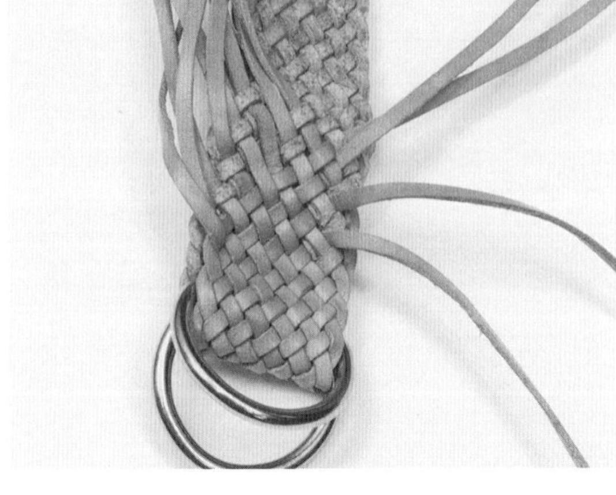

e. Take the lower end on the left, put it under the next end, and insert it under just the second strand from the belt end but not under the strand in the body of the belt, over the next doubled strand, and under the next doubled strand.

f. Take the next, now lower, end on the left, insert it under the first strand at the edge, over the next doubled strand, under just the next strand from the belt end but not the body of the belt, and under the next strand. The strands may be pulled snug, but not so tight as to distort the braid, and trimmed off to ⅛ inch.

5. Make two four-strand keepers, using two ⅛-inch-wide, 16-inch-long strands for each, middled. Braid the keepers firmly to about 2½ inches, stretch them, and place them around the belt. Back-braid one keeper around the belt snugly just behind the rings and the other snugly over the doubled end of the belt. See figure 16-5.

Fig. 16-5. Keeper

Shellac to finish, using a little extra shellac on the back where the rings and keepers are back-braided.

FURTHER SUGGESTIONS

Sixteen-strand belts, 1½ inches wide, are made in a similar fashion. With these the strands are turned starting at about 10 inches. Six-strand keepers are more attractive than four-strand on these wider belts. In Australia, headstalls and reins made in a flat braid of kangaroo were common. Flat braids can be widely used to replace solid strap work, as in camera straps, bag straps, and shoulder straps. Since kangaroo leather is about four times as strong as cowhide, the braided straps are much stronger than the equivalent solid strap.

References and Sources

Edwards, Ron. *How to Make Whips*. Reprint: Cornell Maritime Press, a Division of Schiffer Publishing, Ltd.

A Grant, Bruce. *Encyclopedia of Rawhide and Leather Braiding*. Cornell Maritime Press, a Division of Schiffer Publishing, Ltd. (This carries an extensive bibliography of braiding.)

Morgan, David W. *Whips and Whipmaking*. Cornell Maritime Press, a Division of Schiffer Publishing, Ltd.

Those interested in the types of braiding used in Argentina may consult the following:

Flores, Luis Alberto. *El Guasquero—Trenzados Criollos*. Buenos Aires: Cesarini Hermanos Editores, 1960.

Lopez Osornio, Mario A. *Trenzas Gauchas*, Buenos Aires: Talleres Graiicos Bartolomé U. Chiesino, 1943.

Sources of Supply

Materials and tools are available from local leather shops or from The Leather Factory, P.O. Box 50429, Fort Worth, TX 76105.

Kangaroo skins in braiding tannage are available from David Morgan, 11812 North Creek Parkway N., Suite 103, Bothell, WA 98011. Web site: www.davidmorgan.com